CompLife 2007

To learn more about AIP Conference Proceedings,
including the Conference Proceedings Series, please visit the webpage
http://proceedings.aip.org/proceedings

CompLife 2007

3rd International Symposium
on Computational Life Science

Utrecht, The Netherlands 4 – 5 October 2007

EDITORS
Arno Siebes
Utrecht University
Utrecht, The Netherlands
Michael Berthold
University of Konstanz
Konstanz, Germany
Robert Glen
University of Cambridge
Cambridge, United Kingdom
Ad Feelders
Utrecht University
Utrecht, The Netherlands

All papers have been peer reviewed.

AMERICAN INSTITUTE OF PHYSICS

Melville, New York, 2007
AIP CONFERENCE PROCEEDINGS ■ 940

Editors

A. P. J. M. Siebes
Utrecht University
Department of Information
and Computing Sciences
Padualaan 14, De Uithof
3584CH Utrecht
The Netherlands
E-mail: arno.siebes@cs.uu.nl

Michael R. Berthold
University of Konstanz
Department of Computer
and Information Science
Box M712
78457 Konstanz, Germany
E-mail: berthold@ieee.org

Robert R. Glen
Unilever Centre
for Molecular Science Informatics
Department of Chemistry
University of Cambridge
Cambridge CB2 1EW
United Kingdom
E-mail: rcg28@cam.ac.uk

A. J. Feelders
Utrecht University
Department of Information
and Computing Sciences
Padualaan 14, De Uithof
3584CH Utrecht
The Netherlands
E-mail: ad@cs.uu.nl

L.C. Catalog Card No. 2007935347
ISBN 978-0-7354-0452-6
ISSN 0094-243X

CD-ROM Available: ISBN 978-0-7354-0453-3

Printed in the United States of America

CONTENTS

DATA ANALYSIS

Preface

There is an increasing appreciation between the different scientific communities in computer science and biology that each has substantial goals in common and much to gain by collaboration on complex problems. Providing a forum for a lively exchange between computer scientists, biologists, and chemists remains our goal. After the successful first two editions of the International Computational Life Science Symposium, held in Konstanz (Germany) and Cambridge (UK) respectively, these proceedings report on the third symposium, held in Utrecht, The Netherlands (October 4-5, 2007).

Although we received fewer submissions than in 2006, the quality of the submissions was again high. We believe we have succeeded in putting together an interesting and high quality program consisting of 15 oral presentations.

Of course, a successful conference relies on many people working hard behind the scenes. We wish to express our sincere thanks to the publicity chairs Thorsten Meinl and Andreas Bender. Special thanks to Heather Fyson and Edith van den Hurk for conference and local organization. Finally, we are grateful to the members of our Program Committee; without their effort it would have been impossible to put together such an invaluable program.

August 2007
Arno Siebes
Michael R. Berthold
Robert C. Glen
Ad Feelders

GENOMICS

Improving Microarray Expressions with Recalibration

Karl Fraser, Zidong Wang, Yongmin Li, Paul Kellam[2] and Xiaohui Liu

[1] School of Information Systems, Computing, and Mathematics
Brunel University, Uxbridge, Middlesex, UB8 3PH, U.K.
[2] Department of Infection
University College London, London, W1T 4JF, U.K.

Abstract. Due to the nature of microarray experiments, gene expression levels across and through slide channels can experience up to 10^3 fold change differences in intensity. Such data variance is caused by 'noise' elements, which can influence final expressions. This paper proposes a simple technique whereby histogram transformations are used to reduce noise artefacts. Akin to a magic eraser (removing the top layer of a surface), the technique attempts to blend pixels associated with gene spots into their background. The identification of pixels is relatively straightforward, but blending them with appropriate values is non-trivial. Once replacement values are determined, the background should be a good approximation of the original. By subtracting this surface from the original, gene spot regions would be more accurate. Experiments were carried out and results compared to "GenePix" a mainstream microarray process and "O'Neill" a microarray specific reconstruction algorithm. Not only was our process shown to be significantly quicker in execution time, it also reduced final expression results while typically generating less variation within gene's.

1 Introduction

With application of complementary Deoxyribonucleic Acid (cDNA) microarray technologies, biologists are able to study all of the genes within an organism to obtain a global view of gene interactions and regulations. This technology has huge potential with respect to obtaining greater understanding of the biological processes. However, the technology is still in the early stages of development with improvements required at key phases of the process: spotting, hybridisation, and scanning. Also, because of the multi-layer nature of the process, digitised microarray image data is often permeated with 'noise' that will propagate through all phases of analysis. Realising the technologies true potential therefore is crucial to obtaining high quality image data, which reflects the underlying biology as closely as possible.

Although there has recently been much work focused on how to detect and eliminate various artefacts and errors from natural image data, progress in the microarray field is slow. Indeed, microarray image improvement has received

CP940, *CompLife 2007, 3rd International Symposium on Computational Life Science*
edited by A. Siebes, M. Berthold, R. Glen, and A. Feelders

relatively little attention [1–3] when compared with advances made in post image analysis work; Normalisation [4–6], Modelling [7] and Clustering [8–10].

In essence, we attempt to rebuild a microarray's background such that biological experiment regions are removed from the image leaving very accurate backgrounds. The background can then be subtracted from the original to yield genes that are more precise. Reconstructed expressions as rendered from these regions are compared to those produced by GenePix [11] and O'Neill *et al.* [2].

The paper is organised as follows. We formalise the problem as it pertains to microarray image data and briefly explain the workings of contemporary approaches in the next section. Then, Section three discusses the approach and highlights the steps involved for analysis. In Section four, we start by describing the data used throughout the work and then detail and evaluate the tests carried out for synthetic and real-world data. Section five summarises our findings and defines future directions.

2 Background

Microarray image analysis techniques typically need knowledge of a given gene spot's approximate central pixel as well as the slide's structural layout. A boundary is defined around the gene spot and background pixels, with the medians of these regions taken to be foreground and background intensities respectively. With subtraction of the background medians from those of the foreground, the result can then be summarised as a $\log_2$ ratio. An example of this can be seen with GenePix which uses a circle of varying diameter, while other techniques use partitioning of pixels by use of a histogram [4, 12] or growing a region from the centre [13, 14] for example. For an extensive comparison of these techniques and more details about their implementation, see Yang *et al.* [1].

To assess the slide background relative to the spot intensities, most methods calculate the difference between the spot intensity and the background noise, often by sampling the gene spots surrounding area. This works well when the assumption holds that there is little local variation between what is behind the gene spot and the surrounding area or in situations where the background is evenly distributed over the slide. However, in many situations this is not the case as highlighted in Fig. 1. In 1 the central gene spot's pixel neighbour intensities change significantly (the strong scratch artefact and blob) and more subtly (blooming of other intensities). This blooming can be seen more readily in the outer regions and although not strictly a problem for these sampling methods, this highlights care must be taken when reconstructing from larger areas.

What is needed is a more specific background determination process that can account for the inherent variation between the gene spot and background regions. Indeed the data suffers from an abundance of problems as could be expected from real-world analysis and would include such issues as noise from missing, inconsistent, and outlier elements. No matter how much care is taken when carrying out an experiment, it is certain that there will be errors within

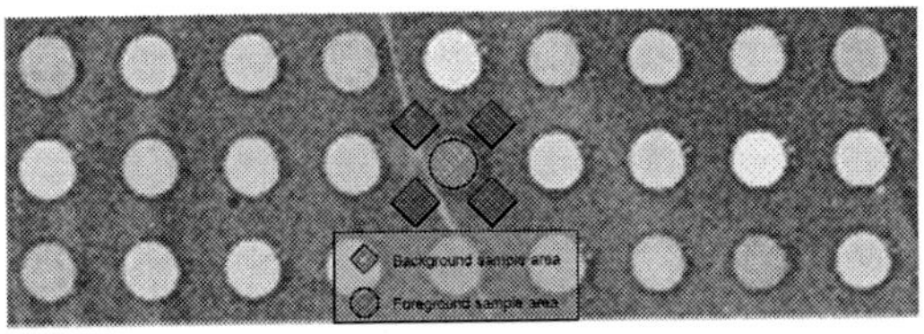

Fig. 1. Sample Gene and Background Locations for GenePix Valleys

the slide. These errors take many forms, but generally, they can be technical, like the random variation in scanning laser intensity, inaccurate measurement of gene expressions and a wide range of artefacts such as hairs or dust on the slide. Alternatively, the errors can be of a more physical nature, inconsistent hybridisation, or washing processes for example.

A possible avenue for such background reconstruction processes is that as represented by the Texture Synthesis community. Efros *et al.* [15] proposed a non-parametric reconstruction technique that is now well established. They worked from the principal of growing an initial seed pixel (located within a region requiring rebuilding) via Markov Random Fields (MRF). Although this works well, the nature of the MRF is such that speed is sacrificed for accuracy. Bertalmio *et al.* [16] on the other hand took an approach relying on the techniques as used by professional restorers of paintings and therefore worked with the principle of an isotropic diffusion model. Chan *et al.* [17] extended these works along with other related techniques and proposed an elastic curvature model approach that combined amongst others Bertalmio's transportation mechanism along with the authors earlier curvature driven diffusion (CDD) work to produce accurate yet relatively slow reconstructions.

Note that all of these techniques were designed to work with natural imagery and therefore provide aesthetic reconstructions. Oliveira *et al.* [18] tried to produce similar results to [16] albeit at a much faster pace. Alas, microarray images on the other hand contain tens of thousands of regions requiring such reconstructions and are therefore computationally expensive to examine with the highlighted techniques. As is typically the case however, something has to give such that significant speed increases can be found and the Oliveira approach is no exception, as we shall see.

With these considerations in mind, O'Neill *et al.* [2] attempted to harness ideas from the highlighted techniques and improve background prediction results while also reducing computation time somewhat. Specifically, O'Neill utilises a simplification of the Efros technique in which gene spots are removed from the surface and recreated by searching known background regions and selecting pixels most similar to the reconstruction border. As the rebuilt region is derived from given border intensities, it is hoped that local background structures will be retained due to an implementation feature of the technique however, only those regions that straddle the centre of an artefact are built most appropriately. The next section describes a technique that attempts to address some of the problems.

3 Proposed Solution

Calibrated Image Reconstruction (CIR) is a simple technique designed to blend gene spot pixels in a microarray image surface into their local background as natural looking and quickly as possible. Although it is the gene spots we are ultimately interested in, their successful blending at this stage will yield more accurate background regions once this new surface is subtracted from the original. CIR therefore, utilises focused histogram calibration at the gene level across the two or more microarray slide channels to determine an accurate estimation of the given gene spots background. Due to the nature of the microarraying process, genes are typically rendered with different shapes and dimensions across channels. Therefore, CIR uses a generic square window Ω^w centred at the gene (as determined by GenePix), to capture all pixels $p_{x,y}$ within a specified distance from this centre.

A gene spot list *srcList* can be defined as *srcList*=$\Omega^w(g_{x,y})$ with Ω^w representing pixels falling into the windowed region only and $(g_{x,y})$ meaning such pixels belong to the gene spot, while list *trgList* denotes background pixels (those not held in *srcList*) and is defined *trgList*=$\Omega^w(\bar{g}_{x,y})$. Although such a blending process is unlikely to create perfect background renditions, it should be possible to improve upon existing template median approach as advocated by packages like GenePix and brute force methods such as O'Neill. Specifically, the pixel lists are blended by histogram calibration or matching. In simple notation a calibration task can be defined

$$s = T(r) \tag{1}$$

where s and r represent the images pixels with grey levels $\in$[0$\sim$1] for the processed and original images over an (x,y) range after some transformation T function. We assume at this point that function T

1. is continuous and r has been normalised to interval [0$\sim$1]
2. that $T(r)$ is single valued and monotonically increasing in interval $0 \leq r \leq 1$
3. and $0 \leq T(r) \leq 1 \mapsto 0 \leq r \leq 1$

Point 2 guarantees that an inverse mapping exists and the intensity range is non-invertible, while point 3 holds that the transformed image will exist in the same dynamic range as the original image. Taking probability density function (PDF) of s, one can transform r into a uniform distribution by using cumulative density function (CDF) as the transforming function.

As a digital image consists (strictly speaking) of discrete values however the probability of grey level r_k occurring within an image can be approximated by summation

$$r_k = \frac{n_k}{n}, k = 0, 1, 2, ..., L - 1 \tag{2}$$

where n denotes total number of input pixels, n_k the number of pixels having grey level r_k and L the total number of possible grey levels in the image. This means discrete PDF and CDF functions can be written

$$s_k = T(r_k) = \sum_{j=0}^{k} r(r_j) = \sum_{j=0}^{k} \frac{n_j}{n}, k = 0, 1, 2, ..., L - 1 \tag{3}$$

$$G(z_k) = \sum_{i=0}^{k} z(z_i) = s_k, k = 0, 1, 2, ..., L - 1 \quad (4)$$

$$z_k = G^{-1}[T(r)], k = 0, 1, 2, ..., L - 1 \quad (5)$$

such that z values will be uniform and independent of r. Briefly, Eqn. 3 maps the original intensity levels into their appropriate levels as based on the original histogram. Eqn. 4 computes transform function G from the histogram as defined by z and Eqn. 5 renders an approximation of the pixel levels for the processed image via the inverse cumulative density function (iCDF) of r. Put another way, calibration is performed by making a frequency count of pixels falling into a given bin location when divided by total pixels (the PDF). The CDF is generated by tracking the accumulation of successive bin contents while the iCDF is created by performing a linear interpolation of the standard CDF function such that samples are evenly spaced in the [0~1] range. In practice, calibration modifies the target pixels (gene spot) distributions into their equivalent range within the source (local background) regions distributions such that the two are more closely matched. Although the process itself is not as sophisticated as the O'Neill technique for instance, the purpose of such work is to show that simple formulations can be highly effective. In addition, such matching has twin advantages of speed of execution (critical to large-scale images) and (more importantly) retention of accurate gene topology. High level pseudo-code for the CIR process is given in Table 1.

4 Experiments

All of the images used in the paper were derived from two experiments conducted using human gen1 clone set [19] data. The human gen1 experiments were designed to contrast the effects of two cancer-inhibiting drugs over different cell lines. The first, **PolyIC**, is classed as the control or normal cell line while the other **Hela** is known as the treatment line. In total, there are 47 distinct slides (with 9216 genes per slide) generated from a series of several specific time points. The gene spots are divided up into distinct blocks such that their demarcation during later analysis is easier for the biologist to perform. Along with the raw microarray slides, there are also the corresponding GenePix processed results.

In order to gain a clear understanding of a given reconstruction events characteristics over the data, we focus on the absolute error of the median expression intensities as calculated from a given gene spots repeat set. However, such expressions are first calculated over a set of 64 synthetic gene spots (SGS) regions, so that performance differences between the proposed (CIR) and comparison (GenePix and O'Neill) techniques can be gleaned more effectively.

4.1 Synthetic gene spot Experiments

Altogether 64 synthetic genes were created in known background areas of a slide with their locations chosen to encompass a range of possible artefacts and 'more

Table 1. Pseudo-Code of CIR Function

```
Input
  nGenes: Total number of gene spots to process
  srcList: List of given gene spot pixels
  trgList: List of given genes background pixels
Output
  outList: srcList pixels recalibrated into trgList pixel range

Function HistogramEstimation(srcList,trgList):outList
For each nGenes
  For each srcList pixel in nGenes
  set nBin to number of pixels in trgList
  calculate trgLists histogram characteristics and set histX
  determine srcLists histogram characteristics and set thisX
  set thisCDF to thiscount accumulation
  set histCDF to histCount accumulation
  scale histogram bins of thisX into histX range
  linear interpolate new set of bins from underlying histBin
  set out to new bin content
  End For
End For
End Function
```

regular' regions as per true genes. Plotting the differences between GenePix calculated backgrounds and the newly reconstructed surfaces yields an understanding of the absolute error per gene (AEPG) for these SGS regions. Fig. 2a presents this AEPG information for the 64 SGS regions.

Note the significant drop in background estimation values via our approach over GenePix (due to GenePix relying on median values). The results of our approach and those of O'Neill are highly similar in their intensity estimates, the differences being related to the handling of artefact regions. The overall mean of the individual process curves are 399, 337, and 339 flux for GenePix, CIR and O'Neill respectively.

If we examine this information while relating results to what is known about the original region, we begin to appreciate the fundamental differences between the approaches. Fig. 2b plots standard deviations of the regions rather than median intensities directly and gives an idea of the smoothness of an SGS area. Overall, there is very little difference between the quality of O'Neill and CIR results (as hinted at by 2a). Notable exceptions are the way in which artefacts have been dealt with during execution. Recall, O'Neill attempts to rebuild the

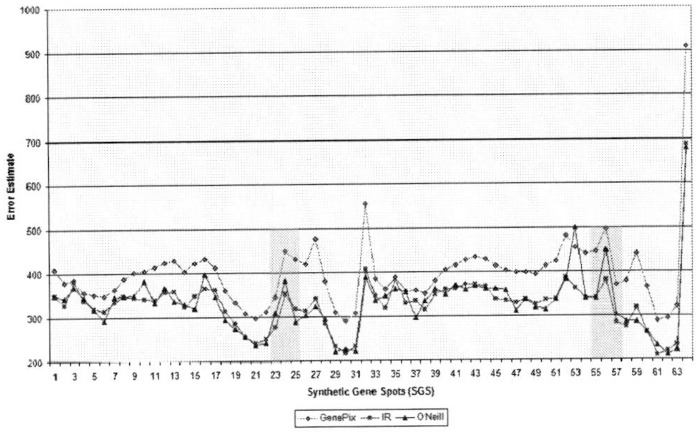

(a)

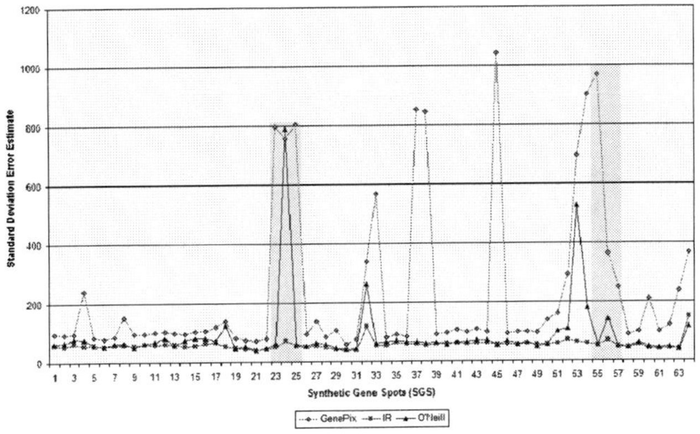

(b)

Fig. 2. Synthetic Gene Spot Curves: Absolute Median (a) and Standard Deviation of (b) Regions

local background and hence remove associated artefact intensity from the region. We, on the other hand take the view that such intensities should be merged with local regions as such artefacts are themselves by-products of DNA binding. Due to their very nature, their removal typically renders the intersecting genes unusable; therefore attempting to filter the artefact will typically help to retain information. The average standard deviation residuals are 100, 55.5, and 62.03 flux for the GenePix, CIR and O'Neill processes respectively which means CIR reduced inter-gene spot standard deviations by half with O'Neill just lagging this result. Crucially, the CIR surface is inherently more accurate as the topology has been retained in a more natural state.

Fig. 3a distils the information from the two previous graphs into their differences by calculating the average absolute pixel error in the SGS regions. In effect, this shows that GenePix generates a potential intensity error of 177 flux per pixel for an SGS region, while the other techniques render smaller estimates. This highlights that downstream analysis (of GenePix results) are usually performed on erroneous gene expressions that background reconstruction can help to improve upon.

4.2 Real gene spot Experiments

With such idealised results detailed and our confidence in CIR enhanced, let us examine characteristics with respect to real gene spot regions. Note that only control genes will be used during this analysis, as there are 768 (24 rows $\times$ 32 columns) of them. The first experiment examines the median intensity of gene spots in one block across a slide to give a feel for typical variances. Fig. 3b presents plots for two reconstruction techniques when compared to the equivalent GenePix performance curve. The reconstruction processes have made subtle changes to the underlying background regions through the image. Other than spike points in the curves the two result sets fluctuate around each other relatively close. Gene 14 and 15 curve spikes are due to these genes becoming saturated during the digitisation process. Note how the two techniques have processed this saturated data differently. O'Neill essentially resets gene 15s background intensity, which translates into subtracting little from the original gene during background correction (BGC). By comparison, CIR estimates a larger background value and would therefore remove more gene intensity during the BGC stage. Clearly, the techniques have different strengths and weaknesses, but the saturation issue reveals a particular inconsistency with the O'Neill process. Genes 14 and 15 were both saturated to a **very similar** intensity level originally and should therefore be processed into a similar result. However, it is clear that O'Neill estimates gene 14s intensity to be much larger than 15 compared with the same genes of CIR, which are more appropriate (their residues alone are three times smaller). Keep in mind however, that as these genes are saturated their topology plateaus and their accuracy should ultimately be questioned regardless. That said CIR has handled these saturated genes more consistently.

Now that we have a feel for general image characteristics, "How do the median characteristics change with respect to a full slide and consequently all slides in the data set?" The second experiment attempts to answer these two points by examining the relationships between expression measurements for all control repeat genes in the one slide initially and then across the entire test set. In all cases the underlying assumption for repeated genes is that they should have highly similar intensity values for a given time point, irrespective of their location on the slide surface. We would expect to see differences as the time point‘s increase over the biological experiments duration.

Fig. 4 represents the absolute foreground medians from both channels for the techniques. Generally, for this particular image, CIR outperformed the other methods comfortably. In addition, we would expect gene intensities to drop with removal of background, as peer background estimation methods tend to over specify this intensity. Fig. 4b presents a cross-sectional average of the 47-slides individual data plots. Although O'Neill outperformed the other reconstruction

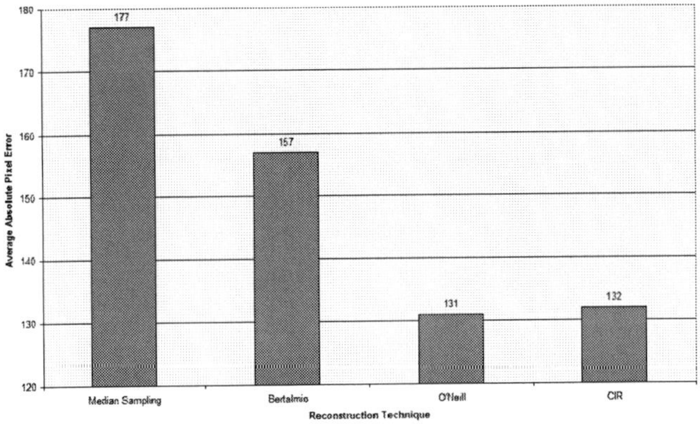

(a)

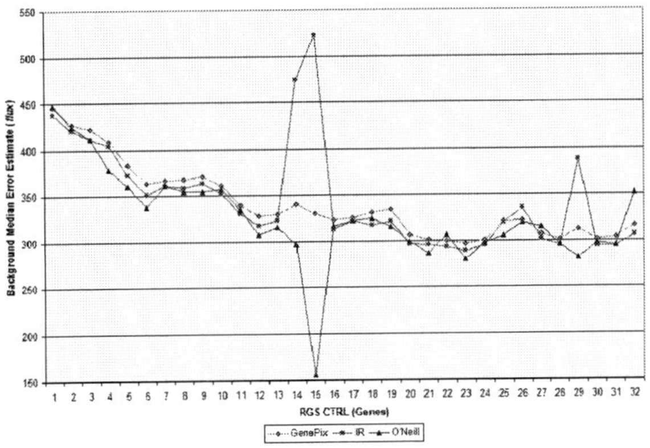

(b)

Fig. 3. Gene Spot Curves: SGS Average Absolute Pixel Error (a), Absolute Medians for 32 genes over block one of a typical slide (b)

methods, note the subtleties of the saturated regions. Clearly, there is a large jump in overall intensity between genes 14 and 15 (again) for O'Neill as opposed to other techniques, but, these genes have very similar intensity (the difference is related to the local region for the gene in question) and should therefore be more similar. Overall, the entire test set yields 10374, 8874, and 9213 flux (for GenePix, CIR and O'Neill).

Clearly this fast matching technique is having a positive effect on the resulting median values, but, "How does this reflect over the test set as a whole?" Fig. 5a attempts to clarify this issue by comparing the improvement (or not) of a particular reconstruction to the original GenePix rendered expressions. In addition, as execution time plays a critical role in reconstruction 5b highlights technique timings for three microarray images. The distinct banding occurring in gene regions 3∼8 and 16∼19 of 5a are associated with saturated (or near)

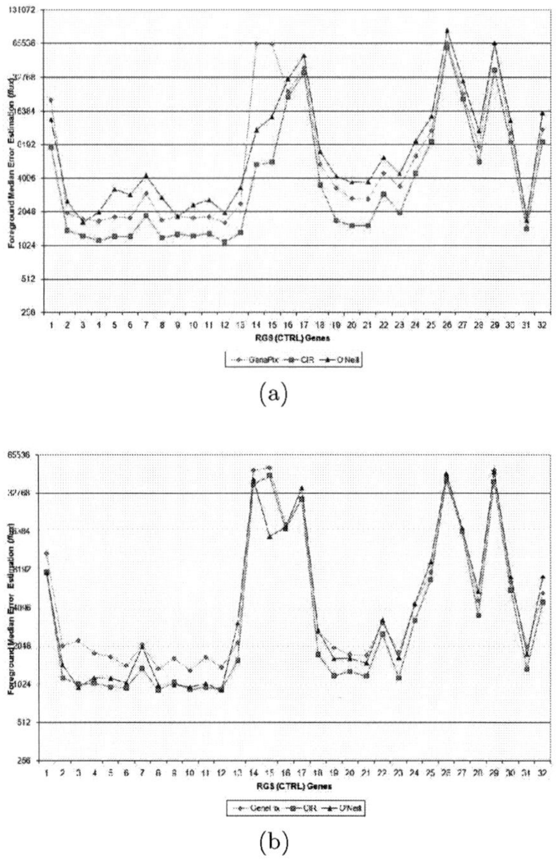

Fig. 4. Gene Spot Curves: Absolute Medians for 32 genes over typical slide (a) and Entire Test set (b) Regions

intensities as created by the scanner hardware. Note that these particular gene groupings represent saturated control genes (they should be saturated). In some cases however, one technique or the other has reduced the repeat variance more, clearly meaning that CIR has more difficulty in this regard.

5 Conclusions

The paper looked at the effects of applying texture synthesis techniques to real-world microarray image data. It has been shown that the use of such methods in medical image applications can be highly effective. We take a different approach to reconstruction with respect to surface topology and provide highly appropriate results, while crucially, reducing computation time for typical imagery significantly.

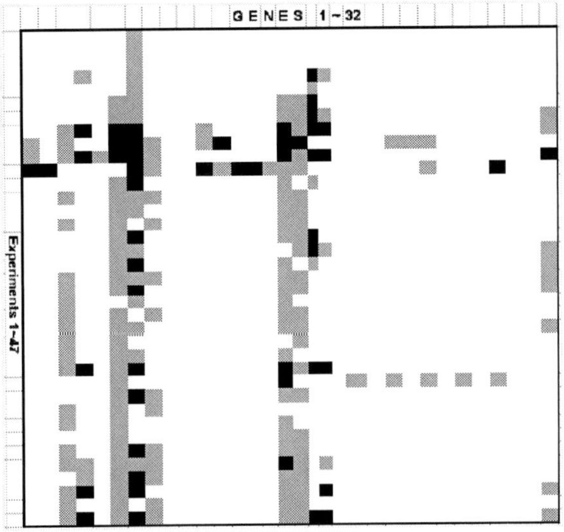

(a)

	Time on Xeon 3.4GHz		
Image	*GenePix*	*O'Neill*	*CIR*
Hela PolyIC t00	4hr+	1hr:53	5mn:47
Hela PolyIC t05	4hr+	1hr:58	6mn:35
Hela PolyIC t18	4hr+	1hr:52	4mn:58

(b)

Fig. 5. Final Results Comparison: Showing difference in repeat expression fluctuations (a); GenePix, CIR and Both techniques are assigned colours black, white and grey (~10% difference) respectively and Sample Timing Chart (b)

The proposed recalibration technique offers a viable step to background reconstruction. Although the formulations themselves are relatively simple, the results show dramatic improvement in background regions. One thing that became quite apparent is the possible advantages of a hybrid process. All gene spots in the images were processed by the given techniques exclusively. However, there will undoubtedly be gene spots that would benefit more from different assimilation methods. Therefore, determining a genes classification could lead to the ability to select the most appropriate reconstruction technique for a given gene. For example, CIR may be most appropriate in a first pass capacity with time intensive techniques better utilised in highly inconsistent regions that fail to satisfy some quality mechanism. Another interesting point is that the current implementation also applies the reconstruction across the whole region equally. Although this clearly works well, in gene spots that cross an artefact rich area, such global consistency may be inappropriate.

Acknowledgment

This work is in part supported by EPSRC grant (EP/C524586/1) and an International Joint Project sponsored by the Royal Society of the U.K. and the National Natural Science Foundation of China.

References

1. Yang, Y.H., Buckley, M.J., Dudoit, S., Speed, T.P.: Comparison of methods for image analysis on cdna microarray data. Journal of Computational and Graphical Statistics **11** (2002) 108–136
2. O'Neill, P., Magoulas, G.D., Liu, X.: Improved processing of microarray data using image reconstruction techniques. IEEE Transactions on Nanobioscience **2**(4) (2003)
3. Bengtsson, A., Bengtsson, H.: Microarray image analysis: background estimation using quantile and morphological filters. BMC Bioinformatics **7**(96) (2006)
4. Chen, Y., Dougherty, E.R., Bittner, M.L.: Ratio-based decisions and the quantitative analysis of cdna microarray images. Journal of Biomedical Optics **2** (1997) 364–374
5. Kepler, B.M., Crosby, L., Morgan, T.K.: Normalization and analysis of dna microarray data by self-consistency and local regression. Genome Biology **3**(7) (2002) research0037.1
6. Quackenbush, J.: Microarray data normalization and transformation. Nature Genetics **32** (2002) 490–495
7. Kellam, P., Liu, X., Martin, N., Orengo, C.A., Swift, S., Tucker, A.: A framework for modelling virus gene expression data. Journal of Intelligent Data Analysis **6**(3) (2002) 265–280
8. Eisen, M.B., Spellman, P.T., Brown, P.O., Botstein, D.: Cluster analysis and display of genome-wide expression patterns. In: Proceedings of the National Academy of Sciences, USA (December 1998) 14863–14868
9. Gasch, A.P., Spellman, P.T., Kao, C.M., Carmel-Harel, O., Eisen, M.B., Storz, G., Botstein, D., Brown, P.O.: Genomic expression program in the response of yeast cells to environmental changes. Molecular biology of the cell **11** (2000) 4241–4257
10. Quackenbush, J.: Computational analysis of microarray analysis. Nature Reviews Genetics **2**(6) (June 2001) 418–427
11. Axon Instruments Inc.: GenePix Pro Array Analysis Software
12. GSI Lumonics.: QuantArray Analysis Software
13. Adams, R., Bischof, L.: Seeded region growing. IEEE Transactions on Pattern Analysis and Machine Intelligence **16** (1994) 641–647
14. Ahuja, N., Rosenfeld, A., Haralick, R.M.: Neighbour gray levels as features in pixel classification. Pattern Recognition **12** (1980) 251–260
15. Efros, A.A., Leung, T.K.: Texture synthesis by non-parametric sampling. In: IEEE International Conference on Computer Vision. (1999) 1033–1038
16. Bertalmio, M., Bertozzi, A., Sapiro, G.: Navier-stokes, fluid dynamics, and image and video inpainting. In: IEEE Computer Vision and Pattern Recognition. (December 2001)
17. Chan, T., Kang, S., Shen, J.: Euler's elastica and curvature based inpaintings. Journal of Applied Mathematics **63**(2) (2002) 564–592

18. Oliveira, M.M., Bowen, B., McKenna, R., Chang, Y.S.: Fast digital image inpainting. In: Proceedings of the Visualization, Imaging and Image Processing, Marbella, Spain (September 2001) 261–266
19. Project., H.G.M.: Human gen1 clone set array

On the Measurement of Genetic Interactions

Edwin de Jong[1], Lude Franke[2], and Arno Siebes[1]

[1] Department of Computer Science
Universiteit Utrecht, The Netherlands
{dejong,arno}@cs.uu.nl
[2] Department of Biomedical Genetics
University Medical Centre, Universiteit Utrecht, The Netherlands

Abstract. Reverse-engineering of genetic interaction networks is one of the primary goals of current bio-informatics research. Accurate knowledge of these networks is of crucial concern for both biological and medical understanding of biological function, and is relevant to the investigation of diseases with genetic origin. In this study, five different measures of genetic interactions are compared with regard to their ability to detect genetic interactions. The results are validated by means of comparison with five databases of known interactions: KEGG, Reactome, BIND, HPRD, and IntAct. A striking finding is that there are substantial differences between the interaction databases with regard to the correspondence with interactions found from gene expression data. These difference overshadow the relatively minor differences between the different interaction measures. These results lead to two conclusions. First, the choice between different interaction measures does not appear do be a crucial factor, given the relatively minor impact of this choice compared to the differences between the results obtained with different interaction databases. Moreover, these findings suggest that gene expression by itself is insufficient as a data source for the detection of genetic interactions. This implies that the identification of other data sources and the integration of multiple data sources are vital issues in improving the quality of bio-informatics approaches to detect genetic interactions.

Keywords: *Gene expression data, genetic interactions, interaction measures, gene interaction networks*

1 Introduction

An important question in current bio-informatics research is how the network of interacting genes can be unravelled based on available data. One of the most important sources of high-throughput data is formed by *Microarrays*; see e.g. [8]. Microarrays are popular because they allow simultaneous measurement of the expression levels of tens of thousands of genes in one or more experimental situations.

Microarray experiments offer (indirect) insight in the molecular processes that take place in a cell. Developing deep insight in the molecular processes in a

CP940, *CompLife 2007, 3rd International Symposium on Computational Life Science*
edited by A. Siebes, M. Berthold, R. Glen, and A. Feelders

cell is a major goal of modern biology. For example, because it would allow the development of drugs for many diseases; ranging from fatal diseases like cancer to non-fatal but incapacitating diseases like celiac disease.

Consequently, the development of techniques that infer interactions from expression data automatically has become an important research topic in Bioinformatics. Because interactions are indicated by correlation of the expression levels, all interaction-prediction methods have to measure correlation. In the data analysis literature, many different correlation measures are known. Hence, a natural question is: which correlation measure is best for interaction prediction? That is the question we address in this paper.

While it is a natural question, it is also a methodologically hard question. First because it is hard, if at all possible, to decompose the quality of a classifier into a part that is due to the correlation measure and a part that is due to the classifier itself. Secondly, which data sets to use? We need both microarray data and known interactions data, for both cases there are many different sources on the web. Thirdly, because for interaction data there is a lack of "negatives". That is, while there are many confirmed interactions, there are preciously few confirmed "non-interactions".

To side-step the first problem, we use the same kind of classifier for all correlation measures. It is a simple threshold classifier to highlight the effect of the different measures as much as possible. For the microarray data, we used the Gene Expression Omnibus and selected data sets with at least 40 experimental situations to make the measured correlations meaningful. For the confirmed interactions, we used five different data sets. The negatives have been defined by an expert. That is, they are determined based on biological knowledge. For example, because their proteins function in different compartments of the cell.

In the next section our methodology is described in detail. The third section presents the results. Section 4 contains the discussion of the experimental results. The conclusions are given in Section 5.

2 Methodology

In this section we detail our methodology. First we introduce the correlation measures. Next we explain our classifier and the "Area Under the Curve" statistic we used to evaluate the different correlation measures. Then we give the details of the databases we used in the experiments. Finally, we explain how we generated negative examples.

2.1 Correlation Measures

As already mentioned in the introduction, there are many different correlation measures. Given that we are interested in measuring the correlation between gene expression levels across different experiments, the different correlation measures can be divided in three groups. The first group contains the parametric measures, in particular those that measure *linear correlation*; see [5] for more background information. Linear correlation has been used among others in Relevance Networks [2].

The second and third group contains non-parametric correlation measures. The difference is that the second group measures *rank correlation*, while the third measures the information theoretic *mutual information..* Mutual information has furthermore been used in the recent ARACNe algorithm [1] and in mutual information relevance networks [3]; see [7] for the original source.

Linear Correlation Probably the best known correlation measure is *Pearson's* r. For two stochastic variables X and Y, it measures in how far there is a *linear relation* between the two. More precisely, r^2 measures how much of the variance of Y is accounted for by the regression of Y on X. On a sample $\{(x_1, y_1), \ldots, (x_n, y_n)\}$ the correlation is given by:

$$r = \frac{\sum_{i=1}^{n}(x_i - \bar{x})(y_i - \bar{y})}{(n-1)s_x s_y}$$

In which, $\bar{x}$ is the sample mean of the x_i and s_x is the sample deviation of the x_i. That is,

$$\bar{x} = \frac{1}{n}\sum_{i=1}^{n} x_i \ ; \quad s_x = \sqrt{\frac{1}{n-1}\sum_{i=1}^{n}(x_i - \bar{x})^2}$$

In our case, the genes are the variables with their expression levels. Since both positive and negative r signal potential interactions, we use the absolute value of r (i.e., $|r|$) as a variant in our experiments.

Rank Correlation While linear correlation looks for linear relations between the two stochastic variables, *rank correlation* looks for monotone relations. More precisely, we first transform the raw data to rankings and then estimate in how far a monotone transformation from one ranking to the other exists.

In our case, each gene puts a ranking on the experimental situations by means of its expression level. The situation with the highest expression level gets the highest rank, the one with the lowest expression level gets the lowest rank. Clearly, it is very well possible that a gene has the same expression level in two or more situations. For such ties, we assign to each situation in the tie the mean of the ranks we could assign to them. So, for example, if gene G has the lowest expression level in situations E_1, E_3, E_{17}, and E_{45}, we assign rank $\frac{1+2+3+4}{4} = 2.5$ to each of these situations.

There are two well-known correlation measures for rankings, *Spearman's* ρ and *Kendall's* τ. We use both of them. Assume we have two rankings R_1 and R_2 of a set of n objects; i.e., experimental situations in our case. Denote by $R_1(o_i)$ the rank that R_1 assigns to object o_i. Moreover, let $d_i = R_1(o_i) - R_2(o_i)$. Spearman's ρ is then defined by:

$$\rho = 1 - 6 \times \sum_{i=1}^{n} \frac{d_i^2}{n(n-1)}$$

Let (o_i, o_j) be a pair of objects with $i \neq j$. Moreover, let *sign* denote the sign-function, which return -1 for negative values, 0 for zero, and 1 for positive values. If for the two rankings R_1 and R_2, we have that

$$sign(R_1(o_i) - R_1(o_j)) = sign(R_2(o_i) - R_2(o_j))$$

the pair (o_i, o_j) is called *concordant*, otherwise it is called *discordant*. Denote by $\#con$ the number of concordant pairs and by $\#dis$ the number of discordant pairs. Kendall's τ is now defined as:

$$\tau = \frac{\#con - \#dis}{\frac{1}{2}n(n-1)}$$

Mutual Information The third class of correlation measures, mutual information, are designed to detect any type of correlation. Different from the previous two, this class originates from Information Theory rather than from Statistics.

Let X and Y, be two discrete stochastic variables with joint probability distribution $P(X, Y)$ and marginal distributions $P(X)$ and $P(Y)$. The mutual information of X and Y, denoted by $I(X; Y)$, is defined by:

$$I(X;Y) = \sum_{x \in X} \sum_{y \in Y} P(x, y) \log \left(\frac{P(x, y)}{P(x)P(y)} \right)$$

Note that as usual in Information Theory, log is to base 2. Computing I from a sample, is slightly more difficult than for our previous correlation measures. It involves the estimation of $P(x), P(y)$, and $P(x, y)$. Especially the latter requires far more data than we will usually have to get a reliable estimate. Therefore we will approximate I by *binning* the data. That is, we partition the data in equal volume bins and estimate the probabilities per bin.

Why So Many? Mutual information is more general than rank correlation, which is in turn more general than linear correlation. Hence, the reader might wonder why we use all three classes. Wouldn't it be sufficient to use only mutual information? Because this will detect any correlation structure detected by the other measures.

While this is true in theory, it is well-known that in practice, each measure is best in detecting the kind of correlation structure it has been designed for. That is, we have a sample with (close to) linear correlation, then a linear correlation measure detects it the clearest.

2.2 Classification and Evaluation

As already noted in the introduction, it is hard (if at all possible) to decompose the accuracy of a classifier into a part that is caused by the correlation measure and a part that is caused by the type of classifier used. One way to circumvent the decomposition problem is use the same type of classifier on top of all the correlation measures.

That leaves the question: which classifier? Our immediate goal is to evaluate the predictive power of the various correlation measures, rather than developing the best classifier. Therefore, we opted for arguably the most basic type of classifier, viz., the *threshold classifier.*

For any $cm \in \{r, \rho, \tau, I\}$, and for any θ in the range of cm, this classifier is defined by:

$$\mathcal{C}^{\theta}_{cm}(G_1, G_2) = \begin{cases} \text{interaction} & \text{if } cm(G_1, G_2) > \theta, \\ \text{no interaction} & \text{otherwise.} \end{cases}$$

Classifiers have long been evaluated on their accuracy only. The higher the accuracy, the better the classifier. Recently, it has been recognised that this simple statistic does not provide sufficient information. The Area Under the ROC Curve (AUC) does provide far more insight; see e.g. [6]. ROC stands for Receiver Operator Characteristic which originated in signal detection theory.

For a given classifier, ROC analysis involves computing the *True Positive Rate* (TPR, or sensitivity) and the *False Positive Rate* (FPR, or 1-specificity). These two are defined by:

$$TPR = \frac{TP}{TP + FN} \text{ and } FPR = \frac{FP}{FP + TN}$$

In which:

TP is the number of true positives, e.g., the number of interactions correctly predicted.

FP is the number of false positives, e.g., the number of interaction incorrectly predicted. In Statistics, such an error is also known as a type I error.

TN is the number of true negatives, e.g., the number of non-interactions correctly predicted.

FN is the number of false negatives, e.g., the number of non-interactions incorrectly predicted. In Statistics, such an error is also known as a type II error.

For a given $cm \in \{r, \rho, \tau, I\}$, each value of θ will give rise to a different (FPR, TPR) pair. We can plot each such point in *ROC space.* That is a plot with FPR on the x-axis and TPR on the y-axis. The curve defined by these points is known as the *ROC-curve.* The area under this curve (AUC) is known to be a good statistic to compare classifiers. Hence, we compare our correlation measures by the AUC they generate.

2.3 Databases

Two types of databases are used for the experiments:

– Microarray data Each of these databases contains gene expression levels, measured using microarrays, for a number of different experimental conditions or subjects, for a number genes. The databases have been selected from the Gene Expression Omnibus (GEO)[1] to satisfy the following conditions:
  • The number of genes in a dataset is at least 1000.
  • The number of measurements in a dataset is at least 40.
– Interaction data In order to validate the interactions found using the differenet measures that are compared, the interactions are compared to databases of known interactions. These databases are curated by biologists, and represent the current state of knowledge regarding which genes interact.

The GEO datasets are listed in table 1, where m denotes the number of measurements. Alla data sets concern human data.

GDS set	#genes	m	description
GDS8	7547	171	Kidney cancer progression (I)
GDS84	6101	84	Breast tumor characterization
GDS89	6313	63	Cancer cell lines (10k_Print3)
GDS181	8295	85	Large-scale analysis of the human transcriptome (HG-U95A)
GDS217	4108	123	Cancer Genome Anatomy Project SAGE library collection
GDS232	1371	46	Medulloblastoma metastasis
GDS330	8295	120	Acute lymphoblastic leukemia treatment responses
GDS531	8295	173	Multiple myeloma and bone lesions
GDS539	8294	63	Chronic granulomatous disease pathophysiology
GDS546	3923	42	Cancer Genome Anatomy Project SAGE library collection
GDS594	1662	158	Large-scale analysis of the human transcriptome (GNF1b)
GDS601	13458	42	T lymphocyte activation gene identification
GDS715	11765	87	Acute myeloid leukemia cell differentiation induced by various drugs
GDS806	13318	60	Estrogen positive breast cancer recurrence during tamoxifen therapy
GDS843	12999	49	Adult acute myeloid leukemia: bone marrow and peripheral blood expression profiles (SHDJ)
GDS913	8295	45	DNA damage from ultraviolet and ionizing radiation effect on peripheral blood lymphocytes
GDS987	11765	41	Kidney transplant response to calcineurin inhibitor-free immunosuppression using sirolimus
GDS1064	11765	43	Acute myeloid leukemia subclasses
GDS1067	11765	52	Plasma cell dyscrasias
GDS1220	11765	54	Malignant pleural mesothelioma

Table 1. GEO datasets used in the comparison

Five different databases of interaction information have been used in the comparison:

KEGG Kyoto Encyclopedia of Genes and Genomes
A biological systems database integrating both molecular building block information and higher-level systemic information. Developed by Kanehisa Laboratories in the Bioinformatics Center of Kyoto University and the Human Genome Center of the University of Tokyo.
URL: http://www.genome.jp/kegg

HPRD Human Protein Reference Database
A centralized platform to visually depict and integrate information pertaining to domain architecture, post-translational modifications, interaction networks and disease association for each protein in the human proteome. All the information in HPRD has been manually extracted from the literature by expert biologists who read, interpret and analyze the published data.
URL: http://www.hprd.org

BIND Biomolecular Interaction Network Database
Descriptions of molecular complexes and pathways and biomolecular interactions. URL: http://www.bind.ca

Reactome A curated knowledgebase of biological pathways
A collaboration among Cold Spring Harbor Laboratory, The European Bioinformatics Institute, and The Gene Ontology Consortium to develop a curated resource of core pathways and reactions in human biology. The information in the database is authored by biological researchers, maintained by the Reactome editorial staff, and cross-referenced with a number of sequence databases. URL: http://www.reactome.org

IntAct A freely available, open source database system and analysis tools for protein interaction data. All interactions are derived from literature curation or direct user submissions and are freely available.
URL: http://www.ebi.ac.uk/intact/site/index.jsf

[1] The GEO datasets, along with related information, can be found at http://www.ncbi.nlm.nih.gov/geo/.

2.4 True Negatives

Defining the True Negative (TN) set is not straightforward [4]. We designed a set of TNs by taking random pairs, but requiring that the eventual degree (number of interacting partners) for each gene, making up interactions, was identical for both the TP set and the TN set, ensuring both networks had an identical (scale-free) topology. This strategy removes a considerable amount of bias, as genes, making up interactions in the TP set will also be present in equal amounts in the TN set. As such, when a gene has not been assessed in a specific GEO dataset, it will not affect the likelihood ratio or AUC.

3 Results

The results of the experiments are shown in both tabulated and graphical form. Tables 2 through 4 provide all results in numerical form. Figure 1 (left) shows the AUC values obtained for the correlation measure against five different interaction databases. Figure 1 (right) through 3 show the average AUC and the standard deviation for all five measures.

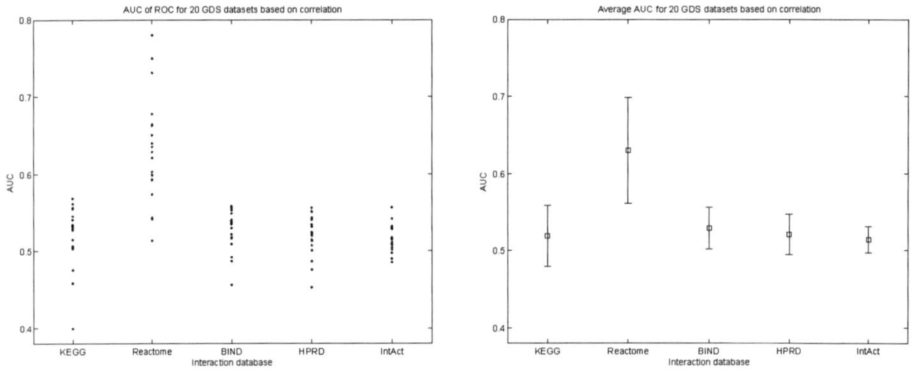

Fig. 1. AUCs for correlation: values (left), and average with errorbars (right).

4 Discussion

First, the differences between the performance of the different measures are small. This means the choice of which measure to use for detecting interactions does not appear to be a crucial factor.

A striking observation is that for all of the first four measures, the performance against Reactome is substantially higher than that against any of the other interaction databases. Even though the variance of the AUC for Reactome

GDS	correl	abscorrel	rankcorrel	tau	mi
8	0.506	0.507	0.475	0.477	0.548
84	0.554	0.510	0.554	0.555	0.530
89	0.533	0.498	0.526	0.525	0.545
181	0.527	0.500	0.528	0.528	0.555
217	0.504	0.521	0.500	0.502	0.531
232	0.545	0.504	0.540	0.548	0.485
330	0.400	0.400	0.445	0.467	0.425
531	0.561	0.485	0.573	0.572	0.503
539	0.530	0.488	0.539	0.538	0.559
546	0.505	0.487	0.501	0.501	0.409
594	0.476	0.451	0.507	0.497	0.467
601	0.515	0.487	0.520	0.522	0.474
715	0.459	0.456	0.457	0.452	0.444
806	0.503	0.499	0.511	0.511	0.505
843	0.529	0.508	0.527	0.527	0.547
913	0.534	0.490	0.536	0.536	0.537
987	0.532	0.504	0.532	0.532	0.496
1064	0.541	0.490	0.539	0.536	0.477
1067	0.568	0.510	0.572	0.568	0.486
1220	0.556	0.511	0.562	0.560	0.493
avg	0.519	0.490	0.522	0.523	0.501
std	0.039	0.027	0.035	0.033	0.044

GDS	correl	abscorrel	rankcorrel	tau	mi
8	0.541	0.540	0.572	0.574	0.421
84	0.678	0.616	0.675	0.675	0.505
89	0.635	0.580	0.624	0.624	0.471
181	0.749	0.693	0.747	0.747	0.470
217	0.629	0.576	0.623	0.621	0.371
232	0.514	0.487	0.521	0.525	0.520
330	0.780	0.780	0.743	0.741	0.702
531	0.663	0.647	0.656	0.657	0.592
539	0.603	0.590	0.602	0.605	0.449
546	0.543	0.506	0.550	0.548	0.530
594	0.593	0.633	0.605	0.606	0.614
601	0.592	0.579	0.593	0.592	0.564
715	0.731	0.731	0.727	0.728	0.713
806	0.598	0.569	0.576	0.578	0.542
843	0.599	0.575	0.608	0.608	0.464
913	0.640	0.594	0.640	0.640	0.415
987	0.573	0.547	0.576	0.576	0.519
1064	0.650	0.614	0.648	0.651	0.562
1067	0.664	0.626	0.662	0.665	0.573
1220	0.621	0.594	0.618	0.620	0.548
avg	0.630	0.604	0.628	0.629	0.527
std	0.069	0.070	0.061	0.061	0.088

Table 2. Results compared against KEGG (left) and Reactome (right).

GDS	correl	abscorrel	rankcorrel	tau	mi
8	0.492	0.492	0.478	0.479	0.508
84	0.557	0.517	0.551	0.551	0.516
89	0.535	0.523	0.535	0.535	0.517
181	0.558	0.516	0.561	0.561	0.518
217	0.555	0.548	0.562	0.562	0.534
232	0.516	0.507	0.529	0.543	0.446
330	0.487	0.487	0.473	0.484	0.472
531	0.557	0.528	0.555	0.554	0.517
539	0.509	0.487	0.513	0.513	0.501
546	0.518	0.529	0.511	0.511	0.506
594	0.456	0.508	0.399	0.387	0.420
601	0.535	0.505	0.535	0.532	0.508
715	0.549	0.550	0.556	0.555	0.542
806	0.540	0.521	0.541	0.541	0.511
843	0.529	0.520	0.531	0.531	0.498
913	0.537	0.510	0.533	0.532	0.503
987	0.521	0.498	0.516	0.516	0.497
1064	0.539	0.515	0.539	0.539	0.498
1067	0.553	0.517	0.558	0.557	0.516
1220	0.539	0.495	0.541	0.541	0.510
avg	0.529	0.514	0.526	0.526	0.502
std	0.027	0.018	0.039	0.040	0.028

GDS	correl	abscorrel	rankcorrel	tau	mi
8	0.487	0.486	0.484	0.484	0.529
84	0.551	0.514	0.546	0.546	0.516
89	0.534	0.503	0.530	0.530	0.512
181	0.543	0.513	0.546	0.546	0.529
217	0.523	0.531	0.525	0.526	0.544
232	0.520	0.505	0.517	0.522	0.477
330	0.453	0.453	0.465	0.479	0.456
531	0.556	0.498	0.562	0.562	0.508
539	0.513	0.495	0.519	0.519	0.517
546	0.507	0.512	0.505	0.505	0.480
594	0.476	0.467	0.489	0.487	0.510
601	0.514	0.497	0.516	0.517	0.498
715	0.501	0.500	0.506	0.503	0.493
806	0.522	0.519	0.526	0.526	0.508
843	0.524	0.514	0.525	0.525	0.516
913	0.540	0.501	0.539	0.538	0.512
987	0.524	0.510	0.518	0.518	0.498
1064	0.531	0.502	0.532	0.531	0.490
1067	0.551	0.504	0.554	0.553	0.497
1220	0.541	0.502	0.547	0.547	0.499
avg	0.521	0.501	0.523	0.523	0.505
std	0.026	0.017	0.024	0.023	0.020

Table 3. Results compared against BIND (left) and HPRD (right).

GDS	correl	abscorrel	rankcorrel	tau	mi
8	0.490	0.490	0.520	0.518	0.494
84	0.542	0.506	0.536	0.536	0.495
89	0.508	0.503	0.507	0.507	0.480
181	0.528	0.528	0.526	0.525	0.490
217	0.517	0.504	0.513	0.514	0.509
232	0.497	0.501	0.514	0.519	0.453
330	0.532	0.532	0.512	0.509	0.510
531	0.510	0.511	0.508	0.508	0.505
539	0.486	0.497	0.488	0.489	0.488
546	0.511	0.487	0.511	0.512	0.503
594	0.530	0.503	0.513	0.520	0.554
601	0.502	0.501	0.499	0.503	0.525
715	0.556	0.557	0.553	0.554	0.544
806	0.508	0.511	0.509	0.509	0.503
843	0.515	0.517	0.513	0.513	0.490
913	0.516	0.494	0.516	0.517	0.475
987	0.508	0.506	0.508	0.508	0.511
1064	0.505	0.520	0.504	0.505	0.506
1067	0.517	0.510	0.516	0.517	0.516
1220	0.508	0.509	0.506	0.507	0.502
avg	0.514	0.509	0.514	0.515	0.503
std	0.017	0.016	0.013	0.013	0.022

Table 4. Results compared against IntAct.

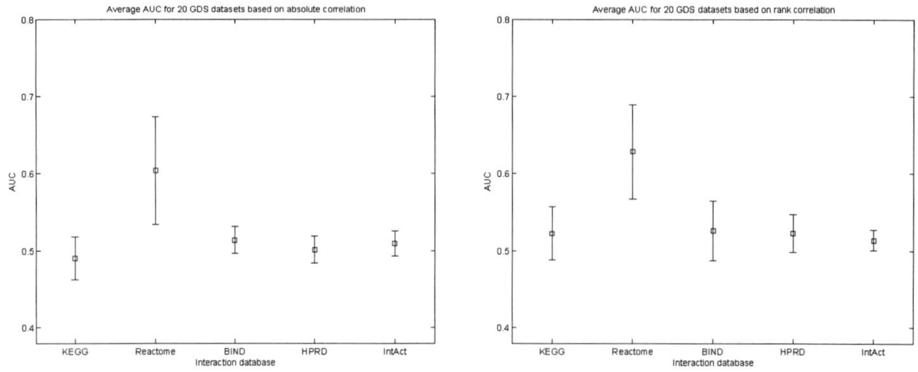

Fig. 2. Average AUC for absolute correlation (left) and rank correlation (right)

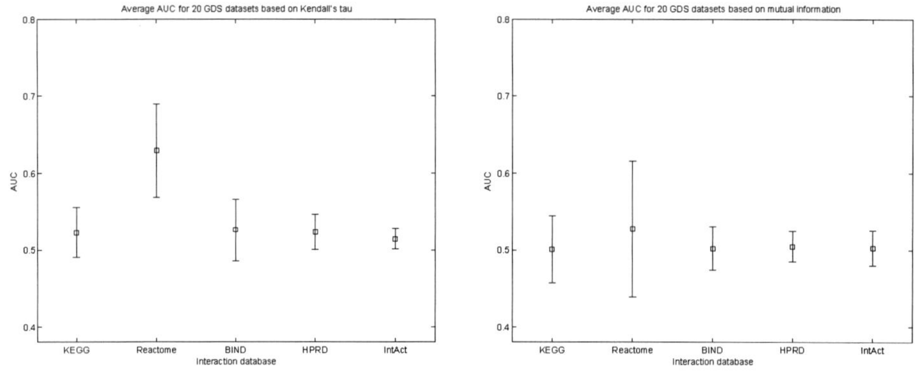

Fig. 3. Average AUC for Kendall's Tau (left) and mutual information (right)

is substantial, the difference is significant for many combinations of measure and interaction database.

A second, related observation is that the AUC is very low for the remaining four interaction databases. While for Reactome average AUCs of over 0.62 are achieved, the average performance against these interaction databases never exceeds 0.53, which is only slightly above chance level (0.5). This suggests that the gene expression data does not provide sufficient information to detect the interactions in these interaction databases.

The results for mutual information differ from the other four measures. For mutual information, the average AUC for Reactome is also low (0.527), though the standard deviation is high (0.088), and values of above 0.70 were also found for mutual information. This reinforces the picture that the interactions in Reactome are more readily found than those in the remaining four interaction databases.

5 Conclusions

The goal of this study was to compare the ability of five different interaction measures to identify which genes interact. A surprising finding is that the differences between the different interaction measures that have been compared are small compared to the differences seen between the data sources used for validation, i.e. the databases of known interactions. Specifically, it was found that for all measures, the interactions found correspond much better to the interactions in the Reactome databases than to those in any of the other four interaction databases employed. A practical consequence is that the choice of measures used to detect interactions does not form a crucial bottleneck, although the mutual information yields a lower performance than the four other interaction measures that have been compared.

Apparently, there is only a weak match between the gene expression data and the interaction databases that have been employed. If this explanation is correct, a promising direction for future research is to identify further data sources that can be employed in order to identify genetic interactions.

Remaining questions of interest are: Why are so few of the known interactions detectable using gene expression data, and: which data sources are most adequate for finding these known interactions? With regard to the latter question, protein interaction data is a primary candidate for an additional data source to augment gene expression data.

References

1. K Basso, AA Margolin, G Stolovitzky, U Klein, R Dalla-Favera, and A Califano. Reverse engineering of regulatory networks in human b cells. *Nature Genetics*, 37(4):382–390, 2005.
2. AJ Butte and IS Kohane. Unsupervised knowledge discovery in medical databases using relevance networks. In *Proceedings of the AMIA Symposium*, pages 711–715, 1999.
3. AJ Butte and IS Kohane. Mutual information relevance networks: functional genomic clustering using pairwise entropy measurements. In *Proceedings of the Pacific Symposium on Biocomputing*, 2000.
4. L. Franke, H. van Bakel, L. Fokkens, E.D. De Jong, M. Egmont-Petersen, and C. Wijmenga. Gene networks and gene prioritization. *American Journal of Human Genetics*, 78:1011–1025, 2006.
5. M. Kendall and A. Stuart. *The Advanced Theory of Statistics*, volume 1. MacMillan Publishing, New York, 1977.
6. Foster Provost and Tom Fawcett. Robust classification for imprecise environments. *Machine Learning*, 42(3):203, 2001.
7. C.E. Shannon and W. Weaver. *The mathematical theory of communication*. University of Illinois, Urbana, 1949.
8. Ernst Wit and John McClure. *Statistics for Microarrays: Design, Analysis and Inference*. Wiley, 2004.

Cluster Assisted Mixed effect Models (CAMM) for the identification of differentially expressed genes

Paul S. McCormick *, Peter Murray**, Robert C. Glen ***

Unilever Centre for Molecular Science Informatics, Department of Chemistry, University of Cambridge, Cambridge, CB2 1EW, UK

Abstract

With many of the major technological issues overcome, the remaining hurdle to the widespread use of microarrays is analytical. With so many genes and often very few replicates, most commonly used statistical techniques are inadequate to the task. Here we present a new technique for identifying differentially expressed genes based on linear mixed effect models and cluster analysis, that combines the advantages of univariate and multivariate methods whilst beginning to take into account the correlation structure of the data. Our results with the Golub leukemia data set have been promising, with our method selecting not only the commonly identified important genes, but several other biologically significant genes as well.
Contact:paul.s.mccormick@gmail.com

Keywords

Microarray, gene expression, cluster analysis, mixed effect models

Introduction

Microarrays have been called by many 'the microchip of the 21st century' [1]. Able to study the expression levels of thousands of genes simultaneously, they are revolutionising our understanding of the genome and all its complex interactions. Whilst the technology was initially limited by cost and accuracy, improvements such as the use of glass as a substrate and improved labelling methods, mean that although there is still work to be done, most of the major technological issues have now been dealt with. The remaining hurdle to the widespread use of microarrays lies in how to analyse the data: with so many genes on a single array and with often very few replicates, most commonly used statistical techniques are inadequate to the task.

Analysis of microarray data can essentially be broken down into five categories: background correction, normalisation, differential expression, classification and gene dependency. The first two of these are involved in the pre-processing of the data, removing noise and re-scaling so that different arrays or dye channels can be compared. The last three categories are the core of the analysis: identifying which genes are differentially expressed, which genes can best be used to classify new samples and how the genes effect each other through a colossal web of interactions.

Due to the complexity of gene expression, it seems prudent to fully understand the simplest of questions, namely which genes are differentially expressed, before

* Illumina Digital, 8 Canham Mews, Canham Road, London, W3 7SR. To whom correspondence should be addressed

** Unilever Colworth, Sharnbrook, Beds, MK44 1LQ, UK

*** Unilever Centre for Molecular Science Informatics, Department of Chemistry, University of Cambridge, Cambridge, CB2 1EW, UK

CP940, *CompLife 2007, 3rd International Symposium on Computational Life Science*
edited by A. Siebes, M. Berthold, R. Glen, and A. Feelders

moving onto such problems as modelling all the gene interactions. Much work has been carried out to this effect, with proposed solutions ranging from looking at fold changes and simple t-tests through to regression models [2] and mixture models [3]. One of the most interesting developments is the application of mixed effect models [4,5], which allows the various sources of error in the data to be individually extracted out and studied.

However, there is as yet no method for computing p-values which combines the flexibility of univariate approaches such as the t-test, which allow the variance of each gene to be different, with the global modelling advantages of multivariate approaches, namely extracting error terms in the data caused by, for example, array or spot effects. Whilst there are some methods which account for both the correlation between genes and gene specific variances [6,7] these are based on pre-calculated p-values, as opposed to using the variance and correlation information to estimate the p-values.

There has also been a great deal of study in clustering genes to look for common functionality [8,9]. Such approaches look for similarities between genes, using distance measures to group genes into clusters such that genes in the same cluster are in some sense 'similar', whilst genes in different clusters are 'dissimilar'. Whilst some work has been carried out in the use of mixed effect models to aid in the clustering of genes [10], no attempt has been made to use clustering to aid in the fitting of mixed effect models in identifying differentially expressed genes.

We propose a novel approach to identifying differentially expressed genes that combines not only the flexibility of univariate approaches with the global modelling advantages of multivariate approaches, but also begins to take into account the correlations between genes. Our approach attempts to simplify the structure of the gene relationships so that this information can be incorporated into a mixed effect model. Whilst all the work so far in applying mixed effect models to microarray data has focused on which random effect terms to include and how best to estimate them, we look at the structure of the remaining ϵ noise terms. Traditionally these terms are considered as IID, with the covariance structure being attributed to the random effect terms such as spot or array effects. However, this may not be the most appropriate approach to take in the context of microarray data [11]. We propose to use clustering of the genes to obtain information to alter the covariance matrix of the ϵ error terms in a linear mixed model (LMM) so that a more flexible multivariate model is fitted, in that only genes with similar variances have the same variance in the model and genes that are highly correlated are correlated in the model. For shorthand we refer to our new method as CAMM (Cluster Assisted Mixed effect Model).

Method

We have developed CAMM initially for Affymetrix gene-chip data so that we need not consider dye-bias effects. Let i index the gene, $i = 1,...p$; j index the array, $j = 1,...,n$; and k index the condition e.g. test and control. Let y_{ijk} be the observed intensity value of gene i, on array j and under condition k.

If we were to assume all the noise in the data is attributable to the different genes i.e. $y_{ijk} \sim N(\underline{\mu}_k, \Sigma_k), \Sigma_k = \{\Sigma_{ii'}\}_k$, then the covariance of genes i and i' under condition k would be given by:

$$\Sigma_{ii'} = \frac{\sum_{r=1}^{n}(y_{irk} - n^{-1}\sum_{s=1}^{n} y_{isk})(y_{i'rk} - n^{-1}\sum_{t=1}^{n} y_{i'tk})}{(n-1)}. \quad (1)$$

Instead, we assume the noise in the data is derived from multiple sources and can be partitioned into noise from the genes and technical noise, such as that caused by using different arrays for different samples. We can take account of such technical noise by utilising a linear mixed effects model (LMM) in which every source of error is assigned a random effect term e.g. $array_j$ would describe noise that effects every gene on an array in the same way but can vary from array to array. As a result of partitioning the noise in the data, the covariance of two genes will no longer be given by equation 1 and must be estimated during the fitting of the LMM. However, the degrees of freedom present in the data ($p * n$ for each condition) are insufficient to estimate a separate covariance term for every entry in the covariance matrix ($\frac{1}{2}p(1+p)$ terms). We must somehow reduce the number of parameters in the model. We choose to estimate a simplified version of the matrix, designed to approximate the most important structural features of the true matrix. To do this we use equation 1 to give an initial estimate of the covariance matrix from which we will elucidate important features. We then use these features to parameterise the covariance matrix in the LMM. In fitting the LMM we will estimate the values of the matrix parameters, obtaining a final estimate of the covariance matrix.

Let Σ_k be the covariance matrix for condition k generated by equation 1 (our initial estimate). We can group together entries in Σ_k that are similar and assign them a common parameter, reducing the total number of parameters that must be estimated. Σ_k is comprised of two different sets of values: the gene variances and the covariances between them. We therefore seek to group the entries of Σ_k in two ways: firstly we shall group together genes with similar variances and assign each group a common variance parameter, reducing the number of different terms on the diagonal of Σ_k; we then group together genes that are highly correlated and assign then a common correlation parameter, effectively reducing the number of different off-diagonal parameters in Σ_k. We group the genes using cluster analysis [12], using two different procedures for the two different types of data (variances and correlations).

We consider the vectors $V_k = \{v_{1,k}, ..., v_{p,k}\}$ where $v_{i,k}$ is the variance of gene i under condition k. Since this data is univariate and we are seeking to group together genes with variances values close together, we calculate a distance matrix for each V_k using the Euclidian distance measure:

$$dist(v_{r,k}, v_{s,k}) = +\sqrt{((v_{r,k} - v_{s,k})^2)}. \tag{2}$$

We subsequently use these matrices in the agglomerative hierarchical cluster analysis method of average linkage. In this way, for each condition we can obtain groups of genes with similar variances.

Since we now wish to group together the genes that are highly correlated within each condition, it is logical to use the Pearson correlation measure to obtain our similarity matrix i.e. equation 1 giving Σ_k. We again use average linkage to cluster the genes based on this matrix, obtaining groups of genes that are highly correlated with each other within each condition. For both clustering methods we use Mojena's 1st stopping rule [13] with $k = 1.5$ to select the optimal number of clusters.

Let us assume we obtain vc_k clusters based on the gene variances for condition k, $\{vclus_{1,k}, \dots, vclus_{vc_k,k}\}$ and cc_k clusters based on the gene correlations for condition k, $\{cclus_{1,k}, \dots, cclus_{cc_k,k}\}$. By combining this information we can consider that for each condition, every pair of genes belongs to one of four groups:

- The genes have similar variance and are correlated to some extent.
- The genes have similar variance but are not correlated.
- The genes have different variance, but are correlated to some extent.
- The genes have different variance and are not correlated.

We can now use this cluster derived information to fit a linear mixed effect model [14, 15].

Let us now consider the LMM:

$$y_{ijk} = gene_{ik} + array_j + \epsilon_{ijk}, \tag{3}$$

where y_{ijk} is the observed un-logged expression value, $gene_{ik}$ is a fixed effect describing the true mean expression for gene i under condition k, $array_j$ is a random effect describing technical noise and ϵ_{ijk} is a noise term describing the biological variation in the data. The $array_j$ random effect term is included to take into account error in the intensity measurements that may be attributed to the different arrays: for example, the arrays may be under slightly different hybridisation conditions. The $array_j$ term is therefore carrying out a form of normalisation on the data and hence the LMM carries out both normalisation and estimation steps simultaneously. However, this does not preclude the use of other normalisation techniques.

We can use the cluster information we have calculated to parameterise the ϵ_{ijk} noise term in the model so that

$$\begin{array}{l} var(\epsilon_{ijk}) = \sigma^2_{vclus_{s,k}} \\ \text{if } i \text{ is in variance cluster } vclus_{s,k} \end{array} \tag{4}$$

$$cor(\epsilon_{ijk}, \epsilon_{i'j'k'}) = \begin{cases} 0 & if\ j \neq j' \\ 0 & if\ k \neq k' \\ 0 & if\ j = j', k = k', i \neq i', i \text{ and } i' \text{ in} \\ & \quad \text{different correlation clusters} \\ \rho_{cclus_{t,k}} & if\ j = j', k = k', i \neq i', i \text{ and } i' \\ & \quad \text{in correlation cluster } cclus_{t,k} \end{cases} \tag{5}$$

The $\sigma^2_{vclus_{s,k}}$ and $\rho_{cclus_{t,k}}$ terms used in this parameterisation are estimated in the fitting of the LMM.

In this way, for each condition, only those genes with similar variances have the same variance in the model, and genes that are highly correlated are correlated in the model. There is no correlation between genes under different conditions using this model, or between the same gene on different arrays within a condition, since we assume our samples our IID i.e. values from different patients (samples) should not be correlated but the genes within a patient may be. In this way, we are freeing up the standard linear mixed effect model which requires all the fixed effects to have the same variance and only allows correlations based on a known variable e.g. an array effect can be used but the fact that the genes themselves will be correlated with each other cannot be taken into account unless we already have a variable linking said genes.

In theory, we can fit this model using S-plus's 'lme' function and use a contrasts matrix to see if the difference between the estimated true mean values for each gene under two different conditions is significantly different to zero. This is based on the statistic:

$$\text{diff}(i) = \frac{\hat{gene}_{i1} - \hat{gene}_{i2}}{\sqrt{(\hat{\sigma}^2_{i1} + \hat{\sigma}^2_{i2})}} \tag{6}$$

where $\hat{gene}_{i1}$, $\hat{gene}_{i2}$, $\hat{\sigma}^2_{i1}$ and $\hat{\sigma}^2_{i2}$ are the estimated means and variances of the parameters $gene_{i1}$ and $gene_{i2}$ respectively. We compare diff(i) to the t-distribution with degrees of freedom equal to the degrees of freedom remaining after fitting the LMM model.

In application however, CAMM is impractical in this form since the computer memory required to fit $2p$ factor levels for the p genes on two arrays is excessive. Even fitting separate models to each condition will not overcome this. As such, we present a modified version of CAMM which can be applied to arrays of any size although certain constraints must be imposed.

We propose fitting a LMM to each variance cluster $vclus_{r,k}, r = 1, ..., vc_k$ of genes, for each condition. Using Mojena's stopping rule this may still cause memory problems if the number of genes in a cluster is too large. To overcome this, a 'gene limit', $gmax$, can be set which fixes the maximum number of genes that can lie in a cluster. After applying Mojena's stopping rule, if the number of genes in the largest cluster, $cmax$, exceeds $gmax$, we move down the dendogram until we reach a level at which $cmax \leq gmax$. At this point we stop and select the number of clusters at this level. This is the motivation behind using a hierarchical clustering algorithm: moving up and down a dendogram is fast and straightforward; bounding the maximum size of a cluster with non-hierarchical methods is not.

Since we are fitting several models separately, it will not be possible to take into account correlations between genes which lie in different variance clusters. Instead we calculate the correlation clusters within each variance cluster, rather than across all the genes.

We use S-plus's 'lme' routine [16] for fitting the LMM. This routine encounters errors if there is only one element in a correlation cluster and so we move up the relevant correlation clustering dendograms until there are at least two elements within each correlation cluster. We also encounter problems if there is only one element in a variance cluster since we do not have sufficient degrees if freedom to fit the LMM model. Instead, we choose to fit a simple linear model, ignoring any random effects.

We calculate diff(i) for gene i as before, combining the information on parameter estimates from the relevant models. We determine the degrees of freedom for comparing diff(i) to the t-distribution by adding the degrees of freedom remaining in the models required to calculate the parameter estimates for gene i.

Whilst in the following results section we make use of multiple comparisons correction techniques to control the FWER, we do so only to provide a cut-off to make comparisons across methods convenient and not because we feel any particular technique is necessarilyoptimal. CAMM should primarily be viewed as an improved method of ranking the genes according to differential expression.

Results

To test CAMM we have used the well-studied Golub leukemia dataset [17]. We sought to identify genes differentially expressed between patients with acute lymphoblastic leukemia (ALL) and acute myeloid leukemia (AML), and then between the subgroups T-cell ALL (T-ALL) and B-cell ALL (B-ALL). We compare CAMM with three other methods: the t-test, Welch's modified t-test and SAM [18], as well as with the results of other authors who have studied this dataset. We have not included a comparison with the standard linear mixed effect model as in such a model the $gene_{i,k}$ factor terms are forced to have the same variance which is clearly an unreasonable modelling assumption to make. Other authors, e.g. [4], who have used LMM to explore microarray data have not used the technique directly to identify differentially expressed genes. In the case of Wernisch et al. the data is bootstrapped to obtain p-values.

Golub et al. partitioned their leukemia dataset into training and test sets. For our comparisons, we will use the training set which consists of 38 bone marrow samples (19 B-ALL, 8 T-ALL, 11 AML) obtained from acute leukemia patients at the time of diagnosis. RNA prepared from bone marrow mononuclear cells was hybridised to high-density oligonucleotide microarrays, produced by Affymetrix and containing 7129 probe sets (including controls) for 6817 human genes. The observed intensity values have been normalised by the authors so that overall intensities for each sample are equivalent. This is carried out by fitting a linear regression model using genes with "P" (present) calls in all samples. The baseline sample is given a re-scaling factor of one, whilst the other samples are re-scaled through multiplication by the inverse of the slope of the linear regression line.

In carrying out our analysis we have not pre-processed the data any further than the work already done by Golub et al. To correct for multiple comparisons we have chosen to use the Bonferroni correction with a nominal significance level of $\alpha = 0.05$. Hence the individual test significance level being used is $\alpha^* = 0.05/7129 = 7.01x10^{-6}$. We set $gmax = 100$, since this was the largest value that could be used without requiring excessive processing time.

For the ALL against AML comparison, the t-test detects 37 genes with significant expression changes and Welch's modified t-test detects 26, whereas CAMM detects 78 genes. SAM, applied with no threshold fold change value and using a delta of 0.479, identifies 2857 genes. However, since SAM uses a different multiple comparisons method, we only consider the ranking of the genes in comparison with CAMM.

Table 1 lists all the genes identified as significantly expressed by CAMM, as well as the ranking of these genes given by the t-test, Welch's test and SAM. The table gives the rank for each method and identifies whether the gene has greater expression in ALL or AML. Numbers in brackets give the ranking of genes not identified by the t-test or Welch's test as significant. Figure 1 shows a heat map image of the expression levels of the CAMM selected genes across the arrays.

Of the 78 genes identified by CAMM, 32 were used by [17] in their prediction algorithm with 14 of them noted as being linked biologically to either ALL or AML. For example, CD33, which shows higher expression in AML, is solely expressed in Monocytic/myeloid lineage cells and induces apoptosis in AML, where as MB1, which shows higher expression in ALL, is solely expressed in B-cells, with both having been demonstrated to be useful in distinguishing lymphoid from myeloid lineage cells. Cyclin D3, Op18 and MCM3, all of which showed higher expression in ALL, encode proteins critical for S-phase cell cycle progression, with Cyclin D3 in particular showing overexpression in aggressive B cell lymphoma, glioblastoma and gliosarcoma. E2A, which shows higher expression in ALL, is linked to a form of pre-B-cell ALL which is characterized by a chromosomal translocation t(1;19)(Q23;P13.3) whilst HOXA9, which shows higher expression in AML, is linked to a form of AML which is characterized by a chromosomal translocation t(7;11)(P15;P15).

However, of more interest were those genes identified by CAMM which were not selected by [17], [2] or [3], all of whom have studied the Golub data set with various methods for identifying differentially expressed genes. In this way we can see if CAMM reveals hitherto unknown information in the data. Several of these newly selected genes have been linked to ALL or AML. Work by [19] suggests that abnormalities of DNA mismatch repair due to defective MSH2, which we identified as having lower expression in AML, could play a key role in leukemogenesis, in particular in AML arising in elderly patients or secondary to previous chemotherapy. The occurrence of abnormalities in rb1, which we identified as having significantly lower expression in AML, in a large subset of various malignancies suggests an important role for this tumour suppressor gene in carcinogenesis, but this varies considerably from one tumour type to another and results in patients with acute myeloid leukemia (AML) have been controversial [20].

For T-ALL against B-ALL, CAMM identified 46 genes, compared to 25, 14 and 2859 genes respectively for the t-test, Welch test and SAM (delta=0.123, no fold change threshold) (Table 2). Figure 2 shows a heat map image of the expression levels of the CAMM selected genes across the arrays. Of the 46 genes we identified, many have been biologically linked to either T-ALL or B-ALL, for example: MB1, which showed higher expression in B-ALL, is a B-cell antigen receptor complex associated protein, specifically expressed in B-cells where as ZAP70, which showed higher expression in T-ALL, is associated with the T-cell antigen receptor zeta chain and plays a role in lymphocyte activation. Other examples include CD19, higher in B-ALL, which may be involved in growth regulation of B-cells; CD79B, higher in B-ALL, which is expression specifically in B-cells; GB DEF, higher in T-ALL, which is a T-cell antigen receptor gene and TCF7, higher in T-ALL, which is a transcriptional activator involved in T-cell lymphocyte differentiation.

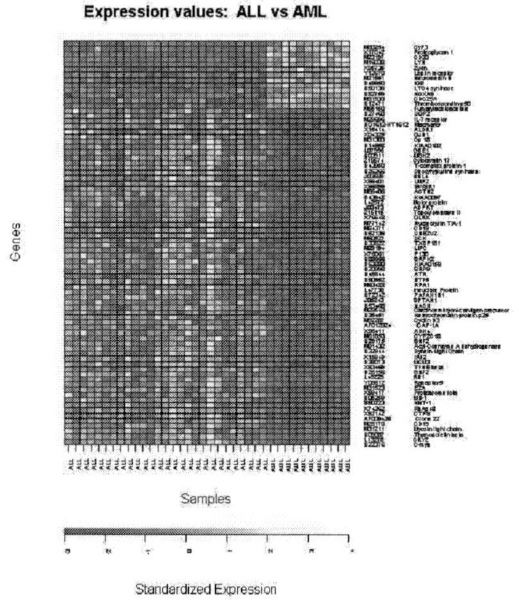

Fig. 1: Standardised expression for 78 genes identified by CAMM as differentially expressed between ALL and AML samples

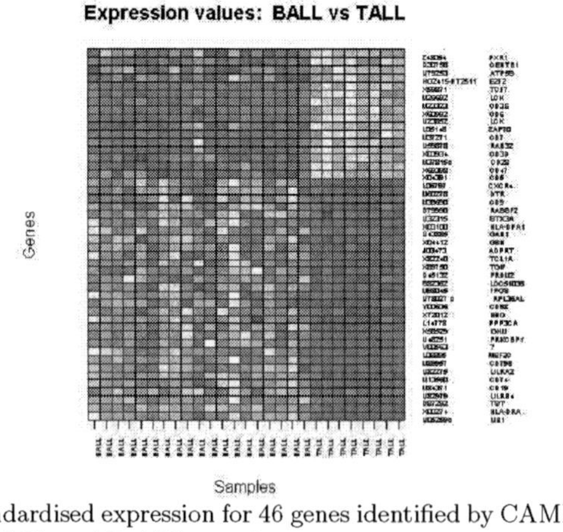

Fig. 2: Standardised expression for 46 genes identified by CAMM as differentially expressed between T-ALL and B-ALL samples

Gene	Probe	CAMM rank	t-test rank	Welch test rank	SAM rank	Higher expression in
Fumarylacetoacetate	M55150	1	3	3	3	AML
C-myb	U22376	2	(40)	1	34	ALL
CRYZ	L13278	3	(96)	5	217	ALL
Thrombospondin-p50	U12471	4	22	21	70	AML
Thymopoietin beta	U09087	5	(139)	8	262	ALL
CDC25A	M81933	6	37	17	72	AML
Myosin light chain	M31211	7	(89)	4	97	ALL
CD19	M28170	8	(157)	11	207	ALL
HOXA9	U82759	9	9	(35)	13	AML
Clone 22	AF009426	10	(147)	12	324	ALL
LTC4 synthase	U50136	11	1	(69)	1	AML
IGIF	D49950	12	7	(93)	31	AML
CTPS	X52142	13	(72)	20	115	ALL
RbAp48	X74262	14	(170)	10	134	ALL
HkrT-1	S50223	15	(120)	13	136	ALL
MB-1	U05259	16	(158)	9	120	ALL
Proteasome iota	X59417	17	(80)	2	54	ALL
E2A	M31523	18	(238)	18	187	ALL
Rabaptin-5	Y08612	19	(233)	23	287	ALL
RB1	L49229	20	(247)	25	314	ALL
SNF2	D26156	21	(90)	6	82	ALL
TFHEβ	X63469	22	(126)	16	173	ALL
Neuromedin B	M21551	23	(53)	(55)	84	AML
MCM3	D38073	24	(153)	15	139	ALL
IRF2	X15949	25	(372)	(43)	380	ALL
Dynein Light Chain	U32944	26	(320)	(30)	225	ALL
Leptin receptor	Y12670	27	5	(155)	5	AML
Acyl-Coenzyme A dehydrogenase	M91432	28	(355)	(40)	265	ALL
SNF2	U29175	29	(330)	(32)	246	ALL
Zyxin	X95735	30	2	(123)	2	AML
CYP2C18	M61853	31	(196)	(28)	483	ALL
ADH4	X56411	32	(123)	(44)	298	ALL
ICAP-1A	AF012024	33	(300)	(41)	572	ALL
Cyclin D3	M92287	34	(138)	7	105	ALL
Heterochromatin protein p25	U35451	35	(306)	(39)	301	ALL
Carcinoembryonic antigen precursor	M29540	36	(173)	(45)	365	ALL
NADH	U53468	37	(191)	(46)	529	ALL
SPTAN1	J05243	38	(207)	22	176	ALL
PAFAH1B1	U72342	39	(237)	(31)	215	ALL
Inducible Protein	L47738	40	(189)	14	151	ALL
RPA1	M63488	41	(370)	(52)	463	ALL
BTF5	U90552	42	(303)	(42)	319	ALL
ATR	U49844	43	(340)	(51)	422	ALL
SRP9	U20998	44	(299)	(27)	214	ALL
KIAA0159	D63880	45	(259)	(36)	244	ALL
SRPK2	U88666	46	(212)	(58)	423	ALL
P130	X76061	47	(290)	(57)	393	ALL
LIPC	M29194	48	(304)	(56)	514	ALL
TXBP181	U33822	49	(152)	(49)	150	ALL
DCK	M60527	50	(413)	(60)	421	ALL
LYN	M16038	51	4	(230)	4	AML
UBE2V2	U62136	52	(472)	(66)	447	ALL
CD19	M84371	53	(384)	(47)	275	ALL
Nucleolysin TIA-1	M77142	54	(501)	(73)	574	ALL
GLRX	X76648	55	(324)	(37)	251	ALL
Topoisomerase II	Z15115	56	(235)	19	171	ALL
ADPRT	J03473	57	(381)	(50)	274	ALL
Motor protein	L42572	58	(419)	(63)	415	ALL
CD33	M23197	59	6	(266)	7	AML
KIAA0097	D43948	60	(385)	(64)	397	ALL
ACTN2	M86406	61	(52)	(142)	66	ALL
EWSR1	X66899	62	(316)	(53)	245	ALL
LMP2	X66401	63	(463)	(65)	363	ALL
UBL4	J03589	64	(244)	(62)	266	ALL
Deoxyhypusine synthase	U26266	65	(140)	(34)	137	ALL
T-complex protein 1	D43950	66	(301)	(59)	242	ALL
Cytokeratin 17	Z19574	67	(222)	(80)	313	ALL
MSH2	U03911	68	(242)	(83)	367	ALL
GBE1	L07956	69	(433)	(75)	405	ALL
KIAA0102	D14658	70	(473)	(74)	387	ALL
Proteoglycan 1	X17042	71	8	(121)	6	AML
Op 18	M31303	72	(133)	24	104	ALL
GJB1	X04325	73	(337)	(84)	392	ALL
ALDR1	X15414	74	(70)	(86)	62	ALL
Macmarks	HG1612-HT1612	75	(203)	26	148	ALL
IL-7 receptor	M29696	76	(454)	(61)	336	ALL
CyP3	M80254	77	13	(271)	21	AML
UGP2	U27460	78	(628)	(99)	516	ALL

Table 1: Genes differentially expressed between ALL and AML samples identified by CAMM, along with rankings for these genes by the t-test, Welch's test and SAM. Numbers give the rankings of the genes by the respective methods; numbers are in brackets when the respective method does not identify the genes as being differentially expressed.

Gene	Probe	CAMM rank	t-test rank	Welch test rank	SAM rank	Higher expression in
CD5	X04391	1	18	4	20	T-ALL
CD47	X69398	2	5	(28)	7	T-ALL
CD28	M37815	3	15	17	26	T-ALL
MB1	U05259	4	(27)	3	23	B-ALL
HLA-DRA	X00274	5	7	1	6	B-ALL
TST	D87292	6	(43)	5	45	B-ALL
LILRB4	U82979	7	(52)	8	62	B-ALL
CD3D	X03934	8	1	11	1	T-ALL
CD19	M84371	9	(58)	7	53	B-ALL
CD74	M13560	10	14	2	13	B-ALL
LILRA2	U82275	11	(83)	12	78	B-ALL
CD79B	M89957	12	(74)	10	63	B-ALL
MEF2C	L08895	13	(129)	(25)	110	B-ALL
Immunoglobulin mu, part of exon 8	V00563	14	(110)	(21)	97	B-ALL
PRKCBP1	U48251	15	(121)	(23)	109	B-ALL
RAB32	U59878	16	(28)	(43)	32	T-ALL
IGHM	X58529	17	(48)	6	43	B-ALL
CD7	M37271	18	4	(75)	4	T-ALL
PPP3CA	L14778	19	(71)	13	71	B-ALL
ZAP70	L05148	20	16	9	14	T-ALL
ENG	X72012	21	(84)	(18)	79	B-ALL
CD58	Y00636	22	(127)	(24)	135	B-ALL
LCK	U23852	23	2	(54)	2	T-ALL
RPL36AL	U78027	24	(119)	22	108	B-ALL
CD6	X60992	25	11	(124)	12	T-ALL
FPGS	M98045	26	(133)	(27)	126	B-ALL
LOC51036	S82362	27	(139)	(32)	279	B-ALL
PRDM2	D45132	28	(141)	(31)	139	B-ALL
CD3E	M23323	29	3	(58)	3	T-ALL
LCK	M26692	30	(13)	(134)	17	T-ALL
TGIF	X89750	31	(179)	(35)	161	B-ALL
TCL1A	X82240	32	(225)	(41)	169	B-ALL
ADPRT	J03473	33	(80)	14	70	B-ALL
TCF7	X59871	34	9	(140)	10	T-ALL
GSN	X04412	35	(142)	(30)	127	B-ALL
E2F2	HG2415-HT2511	36	(75)	(46)	93	T-ALL
GAB1	U43885	37	(160)	(40)	220	B-ALL
HLA-DPA1	X03100	38	(98)	(16)	76	B-ALL
STX3A	U32315	39	(55)	(73)	113	B-ALL
RASSF2	D79990	40	(108)	(26)	104	B-ALL
CD9	M38690	41	(260)	(57)	199	B-ALL
DTR	M60278	42	(171)	(38)	153	B-ALL
CXCR4	L06797	43	(95)	(15)	75	B-ALL
ATP5S	U79253	44	(85)	(48)	90	T-ALL
CENTB1	D30758	45	17	(106)	16	T-ALL
PXR1	Z48054	46	(120)	(45)	132	T-ALL

Table 2: Genes differentially expressed between T-ALL and B-ALL samples identified by CAMM, along with rankings for these genes by the t-test, Welch's test and SAM. Numbers give the rankings of the genes by the respective methods; numbers are in brackets when the respective method does not identify the genes as being differentially expressed.

Discussion

CAMM appears to work well in that it identifies both the important genes commonly selected by other methods, but also several 'new' genes; 'new' in that they have not been previously identified by other authors studying the data set, but have been shown to be potentially biologically linked to the diseases.

As to the method itself, it is by no means in its finished form in that we still plan to expand and test it with block effects and dye effects for cDNA arrays. Overall we view CAMM as a stepping stone, allowing us to begin to model and take into account the dependence structure of the genes which had previously been ignored.

There are obviously many parameters that can be varied throughout the method, such as the clustering method or the value of k in Mojena's stopping rule, which may have an effect on the results. The parameter giving most cause for concern is $gmax$, since this is not based on the data, but rather the computing power available to run CAMM. Hence we have rerun CAMM on the ALL vs. AML data varying the value of gmax between 50 and 200 (larger values cannot be used due to computational limitations). Within this range, the number of differentially expressed genes identified increases with $gmax$ from 68 to 85, with the rate of increase decreasing exponentially as $gmax$ increases. However, the rankings of the genes remains relatively unchanged, with the 68 genes identified with $gmax = 50$ selected for all the $gmax$ values. This suggests that within the range we have investigated, the p-value rankings of the genes are unaltered: only the sizes of the p-values change. This is due to increasing the amount of data used in each of the sub-models: by increasing $gmax$, more genes are present in each model and hence the degrees of freedom are increased, giving us more confidence in the final results and hence smaller p-values. Overall this suggests that whilst more information is gained through increasing $gmax$, relatively small values still give good results. In practise, we would currently recommend running CAMM for various values of $gmax$ and looking for the commonly identified genes.

One potentially problematic issue arising from CAMM is that we are clustering the genes before fitting the model which will, in a sense, normalise the data. To do so we are forced to assume that the normalisation will not effect the data to the extent that genes with similar variances, and hence lying in the same variance cluster, will no longer have similar variances after normalisation. This is an issue we are currently exploring by iterating CAMM, using the estimated variances and correlations to recluster the data and refit the model.

Neither the t-test or Welch's modified t-test identify all of the informative genes described above, even if their results are combined. Whilst SAM selects all the genes identified by CAMM, the ranking of the genes is such that in the ALL against AML case, the top 187 genes in SAM would have to be selected in order to include all of the informative genes selected by CAMM. In particular, the E2A gene, known to be involved in a form of ALL, is ranked 187th. Similarly, for the B-ALL against T-ALL, the top 279 genes in SAM would have to be selected in order to include all of the genes selected by CAMM. In particular, the TCL1 gene for the T-cell leukemia protein 1A is ranked 169th by SAM but 32nd by CAMM, and the CD58 gene, a ligand of the T-cell lymphocyte CD2 glycoprotein, is ranked 135th by SAM but 22nd by CAMM. Similarly, all the CAMM identified genes can also be selected by the t- and Welch tests if the significance value cut-off is set high enough, however, the rankings in Tables 1 and 2 show that if this were done, many other genes would also be selected before all the CAMM identified genes were. Not only does CAMM improve the ranking of important genes, its use of multivariate models appears to increase the degrees of freedom and hence reduce the p-values to the extent that the common problem of the Bonferroni correction being too conservative and failing to identify many or even any genes may no longer occur, though analysis of many other data sets would have to be carried out to confirm this.

Overall then, we have developed a method that appears to perform better than the t-test, Welch test and SAM, identifying both the commonly selected important genes as well as several 'new' genes that are potentially biologically relevant. CAMM goes some way to combining the advantages of both univariate and multivariate methods whilst beginning to take into account the correlation structure of the data. Further work will be to optimise the speed of the CAMM and to extend it to take into account other random effects such as those attributable to using cDNA arrays.

Acknowledgements

This work was funded by Unilever.

References

1. Service, R.: Microchip arrays put dna on the spot. Science **282** (1998) 396–9
2. Thomas, J., Olson, J., Tapscott, S., Zhao, L.: An efficiant and robust statistical modeling approach to discover differentially expressed genes using genomic expression profiles. Genome Research **11** (2001) 1227–36
3. Pan, W.: A comparative review of statistical methods for discovering differentially expressed genes in replicated microarray experiments. Bioinformatics **18**(4) (2002) 546–54
4. Wernisch, L., Kendall, S., Soneji, S., Wietzorrek, A., Parish, T., Hinds, J., Butcher, P., Stoker, N.: Analysis of whole-genome microarray replicates using mixed models. Bioinformatics **19**(1) (2003) 53–61
5. Archer, K., Dumur, C., Joel, S., Ramakrishnan, V.: Assessing quality of hybridized rna in affymetrix genechip experiments using mixed-effect models. Biostatistics **7**(2) (2006) 198–212
6. Dudoit, S., Van der Laan, M., Pollard, K.: Multiple testing. part i. single-step procedures for control of general type i error rates. Statistical Applications in Genetics and Molecular Biology **3**(1) (2004)
7. Korn, E., Troendle, J., Lisa, L., McShane, M., Simon, R.: Controlling the number of false discoveries: application to high-dimensional genomic data. Journal of Statistical Planning and Inference **124**(2) (2004) 379–98
8. Li, H., Hong, F.: Cluster-rasch models for microarray gene expression data. Genome Biology **2**(8) (2001) research0031.1–0031.13
9. De Smet, F., Mathys, J., Marchal, K., Thijs, G., De Moor, B., Moreau, Y.: Adaptive quality-based clustering of gene expression profiles. Bioinformatics **18**(5) (2002) 735–46
10. Ng, S., McLachlan, G., Wang, K., Jones, L., Ng, S.: A mixture model with random-effects components for clustering correlated gene-expression profiles. Bioinformatics **22**(14) (2006) 1745–52
11. Lemieux, S.: Probe-level linear model fitting and mixture modeling results in high accuracy detection of differential gene expression. BMC Bioinformatics **7** (2006) 391
12. Everitt, B., Landau, S., Leese, M.: Cluster Analysis (4th Ed.). Arnold (2001)
13. Mojena, R.: Hierarchical grouping methods and stopping rules: An evaluation. Computer Journal **20**(4) (1977) 359–63
14. McCulloch, C., Searle, S.: Generalized, Linear and Mixed Models. Wiley (2001)
15. Everitt, B., Dunn, G.: Applied Multivariate Data Analysis. Arnold (2001)
16. Pinheiro, J., Bates, D.: Mixed-effects models in S and S-PLUS. Springer (2000)

17. Golub, T., Slonim, D., Tamayo, P., Huard, C., Gaasenbeek, M., Mesirov, J., Coller, H., Loh, M., Downing, J., Caligiuri, M., Bloomfield, C., Lander, E.: Molecular classification of cancer: class discovery and class predicition by gene expression monitoring. Science **286** (1999) 531–7
18. Tusher, V., Tibshirani, R., Chu, G.: Significance analysis of microarrays applied to the ionizing radiation response. Proc Natl Acad Sci **98**(9) (2001) 5116–21
19. Zhu, Y., Das-Gupta, E., Russell, N.: Microsatellite instability and p53 mutations are associated with abnormal expression of the msh2 gene in adult acute leukemia. Blood **94**(2) (1999) 733–40
20. Jamal, R., Gale, R., Shaun, N., Thomas, B., Wheatley, K., Linch, D.: The retinoblastoma gene (rb1) in acute myeloid leukaemia: analysis of gene rearrangements, protein expression and comparison of disease outcome. Br.J.Haematol. **94**(2) (1996) 342–51

Consensus gene regulatory networks: combining multiple microarray gene expression datasets

Emma Peeling and Allan Tucker

Centre for Intelligent Data Analysis
School of Information Systems, Computing and Mathematics
Brunel University
Uxbridge UB8 3PH, UK
emma.peeling@brunel.ac.uk

Abstract. In this paper we present a method for modelling gene regulatory networks by forming a consensus Bayesian network model from multiple microarray gene expression datasets. Our method is based on combining Bayesian network graph topologies and does not require any special pre-processing of the datasets, such as re-normalisation. We evaluate our method on a synthetic regulatory network and part of the yeast heat-shock response regulatory network using publicly available yeast microarray datasets. Results are promising; the consensus networks formed provide a broader view of the potential underlying network, obtaining an increased true positive rate over networks constructed from a single data source.

1 Introduction

Gene Regulatory Networks (GRNs) describe how the expression level of genes effect the expression of other genes. Modelling these networks is a topic of great interest in current bioinformatics research. In this paper we extend on existing Bayesian network methods for modelling GRNs by using multiple microarray datasets to build a consensus model of gene regulatory interactions.

Microarrays are the major source of data for gene expression levels, allowing the expression of thousands of genes to be measured simultaneously. There are now many publicly available microarray datasets under different experimental conditions. However, a key issue with microarray datasets is often referred to as the *curse of dimensionality*. A single microarray dataset usually contains a large number of genes (hundreds or thousands) but the number of observations is much lower - generally tens or up to a few hundred at the very most. This makes it very difficult to extract reliable regulatory interactions from a single dataset. Our research aims to address this by using multiple microarray datasets in order to build a robust consensus model for inferring gene regulatory interactions.

Microarray datasets often come from different platforms and laboratories, across different research groups [1]. This means that datasets contain different biases and it is difficult, or sometimes impossible, to compare the datasets. Measurement units may vary across the datasets and in order to combine the datasets

CP940, *CompLife 2007, 3rd International Symposium on Computational Life Science*
edited by A. Siebes, M. Berthold, R. Glen, and A. Feelders

for analysis, the measurements need to be re-normalised. A key benefit of our proposed framework is that it does not require any special treatment of the data such as re-normalisation and can be used to combine multiple microarray datasets from different experimental conditions or species.

Our method is based on using existing Bayesian network techniques to build a GRN model for each dataset individually which are then used to form a consensus model. Bayesian networks [2] have become a popular method for computational modelling of GRNs from expression data [3–5] since they are able to represent the network qualitatively (with a network graph) and quantitatively (probability distributions quantify the strength of influences and dependencies between nodes/variables in the network graph) and thus are relatively easy to interpret by non-statisticians (e.g. biologists). Our algorithm forms a consensus network by finding the intersection (common relationships in the variable set) of networks generated from different datasets. Only consistent dependencies will appear in the consensus network - providing the effect of removing noise.

Wang *et al.* [6] are the first (to our knowledge) to address the problem of combining multiple time-series microarray datasets for modelling gene regulation. Whilst the high-level framework - the idea of combining solutions from each dataset into an overall, consistent solution - is similar to our approach, their method is based on linear programming where regulatory networks are represented using non-linear differential equations. In this paper we consider the same yeast sub-network that they use, and are able to identify a greater number of documented gene interactions. Additionally they evaluate their method only on time-series experiments from the same laboratory rather than a more diverse set of experiments.

There has been other research published on combining multiple microarray datasets for other tasks, such as functional classification [7]. This research showed that combining multiple datasets can improve classification, even if the experimental focus or conditions varies over the datasets.

The remainder of the paper is organised as follows. In section 2 we describe our approach in more detail. Sections 3 details our experimental results on synthetic and real yeast gene expression datasets. Finally, in section 4 we discuss our findings and outline directions for future research.

2 Methods

2.1 Bayesian networks to model gene regulation

Bayesian Networks (BNs) are graph-based models of probability distributions that capture properties of conditional independence between variables. A BN consists of two components. The first is a Directed Acyclic Graph (DAG) consisting of links between nodes that represent variables in the domain. The second component is a set of conditional probability distributions associated with each node. If there is a link from node A to another node B, then A is said to be a *parent* of B, and B is a *child* or *descendant* of A. The directed arcs between nodes indicate the existence of direct dependencies between the linked variables, whilst the strengths of these influences are quantified by the conditional probabilities. BNs have become a popular method for computational modelling of GRNs from expression data since they are relatively easy to interpret by biologists. The expression level of genes are represented by nodes in the network and influences between genes represented by the directed edges.

BNs capture conditional independence relationships amongst variables. It can be shown that each node is conditionally independent of all its non-descendants given its parents [2]. It is important to mention that more than one DAG may represent the same set of conditional independencies - a set of such DAGs belong to the same equivalence class. It has been shown that equivalent graphs agree on the same underlying undirected structure, but the direction of some arcs may vary [8]. Therefore, the equivalence class of a set of conditional independencies can be represented using a partially directed acyclic graph (PDAG), where only some arcs are directed. It is possible to derive the PDAG representing the equivalence class for any DAG using an algorithm derived by Chickering [9].

2.2 Learning Bayesian networks

We use a score-based search method to learn a BN (representing an equivalence class of networks) that represents a GRN from microarray expression data. Since the first research by Friedman *et al.* [3], search-and-score BNs have been used frequently in learning gene networks. This approach performs a search through the space of possible networks and scores each structure. The aim is to identify the network with the maximum score. A variety of search strategies can be used, the simplest being a greedy hill-climb. We use a simulated annealing approach in order to avoid local maxima. The search begins with an empty network. At each stage of the search, networks in the current neighbourhood are found by applying operators such as *add arc*, *remove arc* and *reverse arc* to the current network.

We use the *Bayesian Information Criterion* (BIC) for scoring candidate networks. The BIC function is a combination of the model log-likelihood and a penalty term that favours less complex models - as such it is similar to the minimum description length:

$$BIC = log\ P(\theta|D) - \ 0.5\ k\ log(n)$$

where θ represents the model, D is the data, n is the number of observations (sample size) and k is the number of parameters. $log\ P(\theta|D)$ is the log-likelihood while the term $0.5\ k\ log(n)$ is a penalty term, which specifically penalises more complex models with more parameters. The BIC is good for dealing with small samples of data as is common with microarray data, as the penalty term helps to prevent overfitting.

Algorithm 1 Consensus Bayesian networks

Input: Set of individual networks (PDAG representation of their equivalence class), consensus threshold (between 0 and 1)
for each pair of nodes i,j **do**
  if link appears between i and j in a proportion of individual networks $\geq$ consensus threshold **then**
    include link between i and j in the consensus network
    Assign directionality according to input networks:
    if there is no directional conflict **then**
      Assign the same direction in consensus network
    else
      if there is a majority direction **then**
        Assign the majority direction in consensus network
      else
        leave arc direction 'unknown'
      end if
    end if
    if Consensus network is cyclic **then**
      leave arc between i and j undirected
    end if
  end if
end for
return Consensus network

2.3 Consensus Bayesian networks

The consensus approach is based on combining the network topologies (graph structures) of multiple BNs. Combining BNs has been addressed previously in two main ways - qualitatively and quantitatively. Quantitative combination is based on aggregating the probability distributions in the networks [10] whereas qualitative combination is based on combining the graph structures [11].

We focus on qualitative combination as we are concerned with the dependency structure amongst the genes. We use graph intersection to generate a consensus network. In other words, since we wish to reduce the noise in the individual networks, only links that appear in all, or a certain proportion of the individual solutions are included in the consensus structure.

The method is detailed in Algorithm 1. It takes as input a set of networks learnt on the same set of variables (nodes) for each dataset and a *consensus threshold* between 0 and 1. Initially, each learnt network is converted to the PDAG representing its equivalence class using Chickering's algorithm. Each pair of nodes is considered in turn and a link between them is created if such a link exists in a proportion of the input networks that exceeds the input consensus threshold.

The directionality of an arc is assigned depending on the input networks. If there is no directional conflict in the input networks then the direction/undirection of an arc remains the same in the consensus network. However, if there is conflict, this introduces some uncertainty regarding the arc direction. At present, if there is a clear majority in the input networks regarding arc direction, then the arc is assigned the majority direction in the consensus network. If there is no majority then the link is left as 'unknown direction'. Note that we make a distinction in the consensus network between edges that are undirected and those that are 'unknown'. An edge that is undirected can be reversed, as in equivalent graphs. However uncertainty exists over the direction of an 'unknown' arc, or whether it can be reversed. We flag up 'unknown' links in the graphs by using arcs that are directed both ways, whereas undirected arcs have no arrowheads.

Topological fusion [11] is a similar method for combining the graphical structures of multiple networks using graph union. This means that the final network structure contains all the edges in every network. Since graph union can introduce cycles into the network structure, arc reversal is also used. This is where an arc $A \rightarrow B$ is reversed and then arcs are added between the parent nodes of A to B, and from the parent nodes of B to A. This maintains the underlying relationships between variables under the principle that it preserves the flow of information [12]. The final fused graph contains all arcs (some reversed) and nodes that are in the input DAGs. We make a comparison between the consensus approach and topological fusion in our experimental results.

3 Results

In this section we report on the experiments performed to evaluate the use of the consensus approach on multiple microarray datasets and compare them to the use of a single dataset and topological fusion. Initial experiments were carried out on a set of four datasets for a synthetic network of 13 genes. We then progressed to a real application: a Yeast heat-shock sub-network.

For each application, a network was generated for each of the n datasets using the score-based simulated annealing search algorithm. The individual dataset networks were used to generate consensus networks at different thresholds (from $\frac{1}{n}$ - corresponding to graph union where at least one of the networks contains each link, through to 1 - strict graph intersection where each link must be contained in every dataset) and a topological fusion network (as described in section 2.3).

We compare the performance of each network in comparison to the true network in terms of true/false positive rates in a ROC space. Whilst the synthetic network was fully known, the yeast regulatory network was constructed from listed gene interactions (both confirmed and potential) in the YEASTRACT database [13]. In a ROC space one can view graphically the performance of a classifier by plotting the true positive (TP) rate (the proportion of true interactions that are identified) against the false positive (FP) rate (proportion of incorrectly identified interactions):

$$TPrate = \frac{TP}{TP + FN} \qquad FPrate = \frac{FP}{FP + TN}$$

where a true positive (TP) is a link that is present in both the learnt and true networks. A false positive (FP) is a link that is present in the learnt network but not in the true network. A false negative (FN) is a link that is in the true network but not in the learnt network, whilst a true negative (TN) is a link that is not in the true or learnt network. Note that these online databases are limited to interactions that have been confirmed by biological studies. Therefore the proportion of FP interactions recorded is likely to be higher than in reality and we are most interested in obtaining a high TP rate and a low FN rate - we do not want to 'miss' documented interactions.

3.1 Synthetic network

In this section we describe the application of the consensus algorithm to a collection of four synthetic datasets. The synthetic regulatory network consists of 100 genes as shown in Fig. 1. For this paper, we only consider the interactions between the regulator genes (transcription factors), genes numbered 1-13. Four time-series expression datasets were generated for the network using differential equations to mimic transcriptional behaviour. Each dataset had a varying amount of noise and a large number of samples ranging from 200-600.

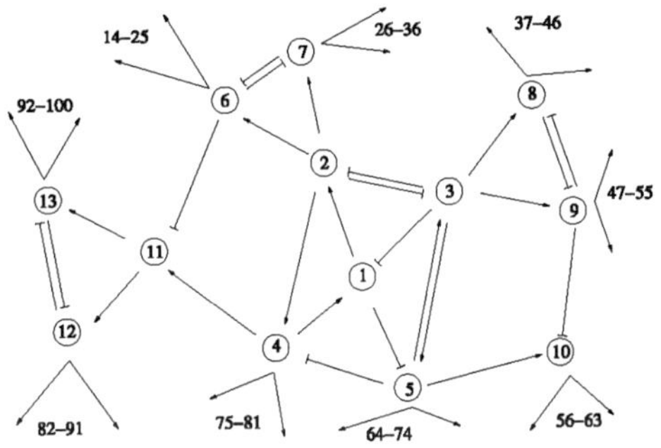

Fig. 1: Synthetic gene regulatory network. Arrows with pointed ends indicate activation, flat ends indicate repression.

The ROC plot comparing the consensus performance against that of individual datasets is shown in Fig. 2. In general, we find that the consensus networks provide a balance between the individual solutions, performing better than datasets 1 and 4, but less well than datasets 2 and 3. The 0.25 consensus network is shown in Fig. 3a. This network raises the TP rate to 89% - capturing almost all links - missing only two known interactions. However, the FP rate is also raised significantly to 36%. Although it is also graph union, the topological fusion network does not do as well as the 0.25 consensus network, as it obtains 6 misdirected arcs. This is because our consensus method represents networks using equivalence classes (so if arcs are reversible they are left undirected) and are therefore less likely to include misdirected arcs.

We can show which interactions are stronger and which are less robust by considering how many individual networks contain each arc. The most robust

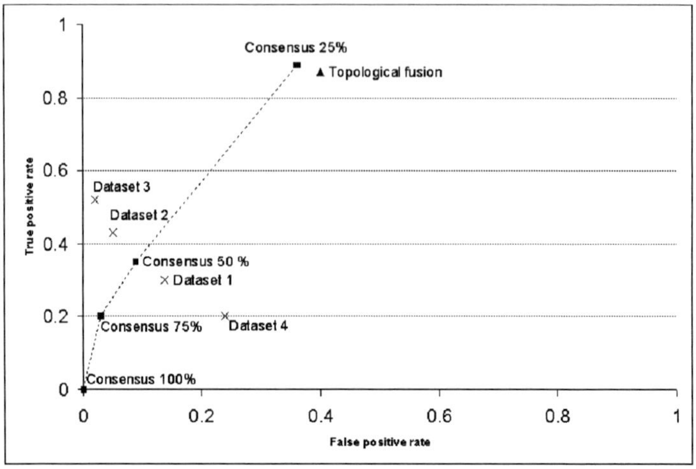

Fig. 2: Comparing the performance of individual dataset and consensus networks to the true network for synthetic data in a ROC space.

Table 1: Similarities between each dataset; measured using link identification accuracy (the percentage of identical links in both networks) on the DAG equivalence classes from their learnt networks.

Dataset	1	2	3	4
1	1	0.85	0.77	0.67
2	0.85	1	0.82	0.69
3	0.77	0.82	1	0.67
4	0.67	0.69	0.67	1

(a) Synthetic network

Dataset	Beissbarth	Eisen	Gasch	Grigull	Spellman
Beissbarth	1	0.44	0.42	0.40	0.40
Eisen	0.44	1	0.51	0.56	0.44
Gasch	0.42	0.51	1	0.49	0.31
Grigull	0.40	0.56	0.49	1	0.44
Spellman	0.40	0.44	0.31	0.44	1

(b) Yeast heat-shock response network

arcs are found in more individual networks. Fig. 3b shows the consensus network (0.25 threshold) with arcs weighted according to their robustness. It can be seen by eye there is a correlation between the more robust links and the TP links.

Table 1a shows a similarity matrix between the individual dataset networks, where similarity is measured using the percentage of arcs that are identical between the networks (we call this link identification accuracy). Dataset 4 is clearly the most different with accuracies of 67-69% whereas the other three datasets obtain accuracies of over 77% between each other. Dataset 4 also performs least well out of all the individual networks in comparison to the true network (see Fig. 2) - indicating that a correlation may exist between dataset similarity and similarity to the true network, and that dataset 4 may be particularly noisy.

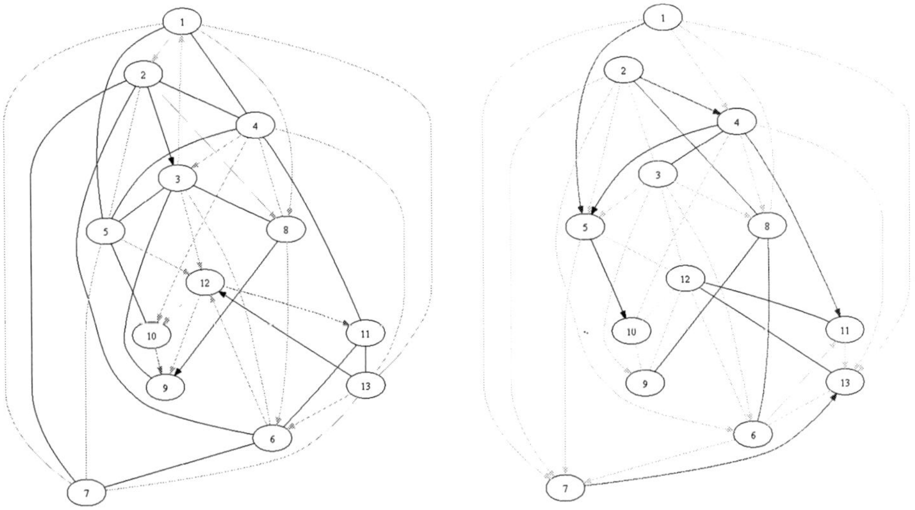

(a) Consensus network (b) with weighted links

Fig. 3: (a) Consensus regulatory network for the synthetic datasets. Correct edges (true positives) are shown in bold black. 'Missed' edges (false negatives) are shown in green, and incorrect edges (false positives) are shown in grey. Red edges represent correct links that are misdirected. (b) Consensus network with weighted arcs - black arcs are found in 75% of individual networks, blue links are found in 50% of networks and lighter blue (cyan) links are found in only one (25%) of individual networks

3.2 Yeast heat-shock response network

We then applied the consensus algorithm to real microarray gene expression data for yeast. We used five different microarray datasets that are publicly available on the YeastBASE expression database - a summary is provided in Table 2. We take the example of 10 transcription factors (TFs) related to heat-shock response from Wang *et al.* [6] in order to evaluate the algorithm on a sub-network of a manageable size and make a direct comparison between the two methods. Two of the TFs selected (HSF1 and SKN7) are known to be directly involved in heat-shock response and are documented as regulating 4 TFs among the 10.

Figure 4 compares the performance of each individual dataset network and the consensus networks in comparison to the true network in a ROC space. The

Table 2: Summary of yeast datasets

Dataset	Description	Observations
Beissbarth [14]	Heat-shock response	12
Eisen [15]	Cold-shock and heat-shock response	14
Gasch [16]	Environmental changes including heat-shock response	173
Grigull [17]	Heat-shock response	27
Spellman [18]	Cell-cycle	73

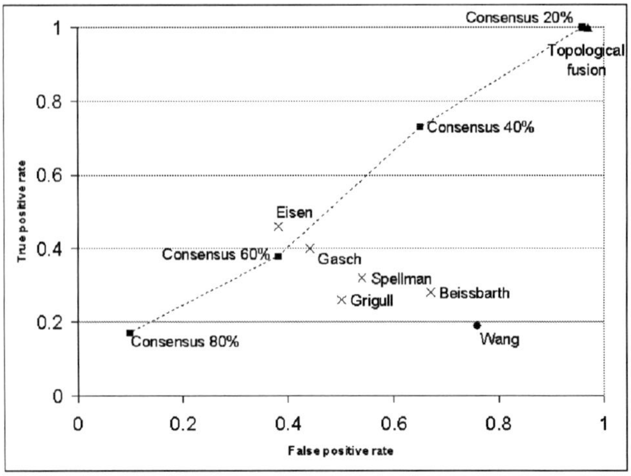

Fig. 4: Comparing the performance of individual dataset and consensus networks to the true network for the yeast heat-shock data in a ROC space

0.4 consensus network (shown in Fig. 5a) obtains the best results, with a TP rate of 73% and an FP rate of 65%. Overall, this network found 10 documented and 6 potential links. It missed 4 actual and 2 potential links and there were 15 incorrect links, 2 of which were actually misdirected documented interactions. In comparison to each individual network it performs as well or better in terms of the proportion of correctly identified links and the TP rate, although it must be noted that the FP rate is higher than 4 of the 5 individual dataset networks. However, some FP links may be valid, but are currently unknown. The topological fusion network obtains a higher FP rate than the 0.2 consensus network (our version of graph union). As with the synthetic data it obtains a higher proportion of misdirected arcs than the 0.2 consensus threshold network because it does not consider the DAG equivalence classes.

The yeast results show a similar pattern to the synthetic data results. On the lower-threshold consensus networks, the TP rate is high, but so is the FP rate. As the threshold increases, both rates drop, but we are able to gain a perspective on which links are more robust. Fig. 5b shows this on the 0.4 consensus network - darker links are stronger, found in more of the individual networks.

Our consensus networks outperforms the network generated by Wang *et al.* [6], which confirms just 3 documented interactions (19% TP rate) with a high number of FP arcs at 76% - although 8 of these are misdirected known interactions. They only use some of the time-series in the Gasch dataset to generate their consensus network, whereas our consensus networks are generated from datasets under various experimental conditions and obtained from different laboratories.

Table 2b shows the similarity between the individual networks. In particular the Eisen and Gasch datasets are most similar whereas the Spellman data is the most different from the others. Like the synthetic datasets, this is reflected in

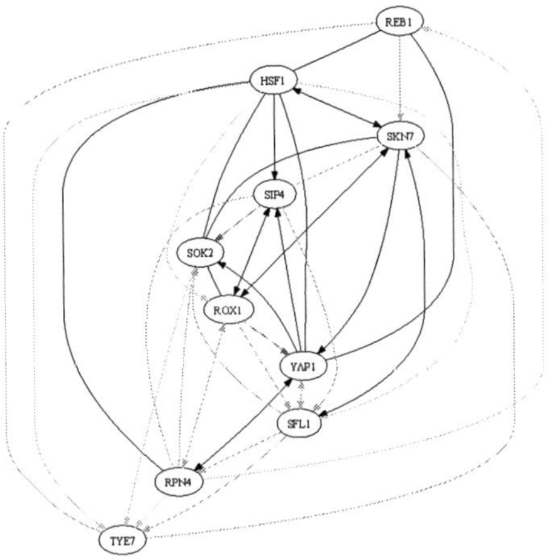

(a) Consensus network

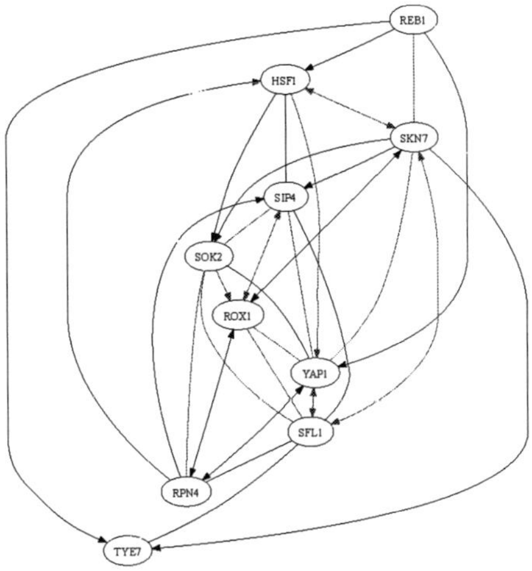

(b) with weighted links

Fig. 5: (a) Consensus regulatory network for the heat-shock response microarray data in yeast. Correct edges (TPs) are shown in bold black. 'Missed' edges (FNs) are shown in green, and incorrect edges (FPs) are shown in grey. Red edges represent correct links that are misdirected. (b) Consensus network with weighted arcs - black arcs are found in at least 2 (40% of) individual networks, and blue arcs in only 1 network.

the comparison with the true network where Eisen and Gasch are closest to the true network.

4 Conclusions

In this paper we have presented an approach for inferring gene regulatory networks from multiple microarray datasets using a consensus Bayesian network approach. In this preliminary research we have evaluated the approach on both synthetic data and real yeast experimental data. Our results show that consensus networks can provide a broader view of the potential underlying network, obtaining an increased TP rate over the individual solutions. The FP rate is also higher - but these network arcs may represent previously unknown but correct interactions.

Our consensus approach provides a process for identifying the more robust links - those gene interactions are identified in more datasets and thus, that are more likely to exist - by varying the consensus threshold. Further work is necessary to produce an automated method for choosing the 'best' consensus threshold. At the moment it is up to the user or an expert to evaluate the output networks.

We also believe that varying the influence of datasets through weighting would improve the approach. Analysis on dataset similarity showed a correlation between the least similar datasets and a lower accuracy to the true network.

Assigning weights to datasets would allow particular datasets to have more influence than others when forming the consensus network. This should improve results by increasing the accuracy of the consensus network closer to the least noisy individual networks.

As well as combining networks from different datasets, the consensus algorithm could be used to combine networks generated from the same dataset, using different methods. This concept is based on the idea of consensus clustering [19]. Different BN learning techniques such as constraint-based learning (e.g. the PC algorithm) and search-and-score techniques, and/or continuous and discretised versions of the same datasets could be combined to produce a consensus network.

Since our approach is based on combining networks, it has the potential to integrate many heterogeneous types of data - provided that regulatory models can be built from these datasets.Future steps in this research project include the incorporation of other data sources or expert knowledge such as transcription factor binding sites, the gene function ontology and textual information extracted from scientific literature. Additional further work as part of the project will also involve extending the modelling techniques to incorporate temporal information through time nodes and dynamic Bayesian networks (this should improve the directionality of learnt interactions and allow cyclic behaviour to be introduced), and the use of hidden nodes to model unobserved variables.

References

1. Soinov, L.: Supervised classification for gene network reconstruction. Biochemical Society Transactions **31**(6) (2003) 1497–1502
2. Pearl, J.: Probabilistic Reasoning in Intelligent Systems: Networks of Plausible Inference. Morgan Kauffman, San Francisco, CA, USA (1991)
3. Friedman, N., Linial, M., Nachman, I., Pe'er, D.: Using Bayesian networks to analyze expression data. Journal of Computational Biology **7**(3-4) (2000) 601–620
4. Hartemink, A., Gifford, D., Jaakkola, T., Young, R.: Bayesian methods for elucidating genetic regulatory networks. IEEE Intelligent Systems **17**(2) (2002) 37–43
5. Pe'er, D., Tanay, A., Regev, A.: Minreg: A scalable algorithm for learning parsimonious networks in yeast and mammals. J. Machine Learning Res **7** (2006) 167–189
6. Wang, Y., Joshi, T., Zhang, X., Xu, D., Chen, L.: Inferring gene regulatory networks from multiple microarray datasets. Bioinformatics **22**(19) (2006) 2413–2420
7. Ng, S., Tan, S., Sundararajan, V.: On combining multiple microarray studies for improved functional classification by whole-dataset feature selection. Genome Informatics **14** (2003) 44–53
8. Pearl, J., Verma, T.: A theory of inferred causation. In: Proceedings of Knowledge Representation and Reasoning 2, New York, USA, Morgan Kauffman (1991) 441–452
9. Chickering, D.: A transformational characterization of equivalent Bayesian network structures. In: Proceedings of Uncertainty in Artificial Intelligence 11. (1995)
10. Pennock, D., Wellman, D.: Graphical representations of consensus belief. In: Proceedings of Uncertainty in Artifical Intelligence 15, Morgan Kauffman (1999) 531–538

11. Matzkevich, I., Abramson, B.: The topological fusion of Bayes nets. In: Proceedings of Uncertainty in Artificial Intelligence 8, Morgan Kauffman (1992) 191–198
12. Shachter, R.: Evaluating influence diagrams. Operations Research **34**(6) (1986) 871–882
13. Teixeira, M., Monteiro, P., Jain, P., Tenreiro, S., Fernandes, A., Mira, N., Alenquer, M., Freitas, A., Oliveira, A., S-Correia, I.: The YEASTRACT database: a tool for the analysis of transcription regulatory associations in Saccharomyces cerevisiae. Nucleic Acids Research **34** (2006) D446–D451
14. Beissbarth, T., Fellenberg, K., Brors, B., Arribas-Prat, R., Boer, J., Hauser, N., Scheideler, M., Hoheisel, J., Schutz, G., Poustka, A., Vingron, M.: Processing and quality control of dna array hybridization data. Bioinformatics **16**(11) (2000)
15. Eisen, M., Spellman, P., Brown, P., Botstein, D.: Cluster analysis and display of genome-wide expression patterns. PNAS **95**(25) (1998) 14863–8
16. Gasch, A., Spellman, P., Kao, C., Carmel-Harel, O., Eisen, M., Storz, G., Botstein, D., Brown, P.: Genomic expression program in the response of yeast cells to environmental changes. Mol. Cell **11** (2000) 4241–4257
17. Grigull, J., Mnaimneh, S., Pootoolal, J., Robinson, M., Hughes, T.: Genome-wide analysis of mrna stability using transcription inhibitors and microarrays reveals post-transcriptional control of ribosome biogenesis factors. Mol. Cell **24**(12) (2004) 5534–47
18. Spellman, P., Sherlock, G., Zhang, M., Iyer, V., Anders, K., Eisen, M., Brown, P., Botstein, D., Futcher, B.: Comprehensive identification of cell cycle-regulated genes of the yeast saccharomyces cerevisiae by microarray hybridization. Mol. Cell **9** (1998) 3273–3297
19. Swift, S., Tucker, A., Vinciotti, V., Martin, P., Orengo, C., Liu, X., Kellam, P.: Consensus clustering and functional interpretation of gene-expression data. Genome Biology **5** (2004)

Using Sorting by Reversal: Breakpoint Graph for Gene Assembly in Ciliates

Robert Brijder and Hendrik Jan Hoogeboom
Leiden Institute of Advanced Computer Science, Universiteit Leiden,
Niels Bohrweg 1, 2333 CA Leiden, The Netherlands,
rbrijder@liacs.nl

Abstract. The theory of gene assembly in ciliates has much in common with the theory of sorting by reversal. Both model processes that are based on splicing, and have a fixed begin and end product. The main difference is the type of splicing operations used to obtain the end product from the begin product. In this overview paper we show that the concept of breakpoint graph, known from the theory of sorting by reversal, has many uses in the theory of gene assembly. Our aim is to present the material in an intuitive and informal manner to allow for an efficient introduction into the subject.

1 Introduction

Ciliates are single cell organisms that have two functionally different nuclei, one called micronucleus and the other called macronucleus. During sexual reproduction a micronucleus is transformed into a macronucleus in a process called gene assembly. This is the most involved DNA processing in living organisms known today: the genome of the micronucleus may be dramatically different from the genome of the macronucleus.

A better known research area concerned with DNA processing is *sorting by reversal*, see, e.g., [1–3]. Two different species can have several segments in their genome that are very similar, although there relative order (and orientation) may differ in both genomes. In the theory of sorting by reversal one tries to determine the number of reversal operations needed to reorder such a series of genomic 'blocks' from one species into that of another. An essential tool is the *breakpoint graph* (or reality and desire diagram) which is used to capture both the present situation, the genome of the first species, and the desired situation, the genome of the second species.

Motivated by the breakpoint graph, the notion of *reduction graph* was introduced in [4] into the theory of gene assembly. The intuition of 'reality and desire' remains in place: instead of two different species, we deal with two different nuclei — the reality is a gene in its micronuclear form, and desire is the same gene but in its macronuclear form.

This research was supported by the Netherlands Organization for Scientific Research (NWO) project 635.100.006 'VIEWS' within the Computational Life Sciences (CLS) program. We thank Tim Cocx for proofreading the paper.

CP940, *CompLife 2007, 3rd International Symposium on Computational Life Science*
edited by A. Siebes, M. Berthold, R. Glen, and A. Feelders

In contrast to sorting by reversal, in gene assembly, the operations that splice the DNA are irreversible and can only be applied on a fixed pair (or two pairs) of positions in the DNA, called *pointers*. Therefore, although the intuition is the same, the definition of reduction graph is notably different from the definition of breakpoint graph (in the theory of sorting by reversal). Surprisingly, where the breakpoint graph is mostly useful to determine the number of needed operations, the reduction graph has different uses in the theory of gene assembly [4–6] (conference versions appeared in CompLife '05 [7] and '06 [8]). In this paper we will summarize its uses and provide an intuitive foundation. As this is an informal account of the material, formal definitions are sometimes and proofs are always omitted.

2 Gene Assembly in Ciliates

In ciliates, genes can occur in two forms: in their micronuclear and in their macronuclear form. A gene in its micronuclear form consists of relevant genetic segments called MDSs (macronuclear destined sequences) separated by IESs (internally eliminated sequences). During sexual reproduction, a micronucleus is converted into a macronucleus. The genes are converted to their macronuclear form by excising the IESs and by splicing, permutating, and possibly inverting the MDSs. This process is referred to as *gene assembly* [9].

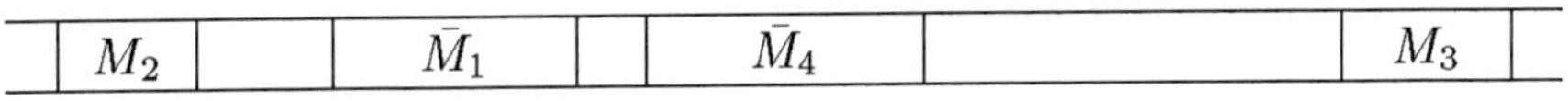

Above is an example gene in its micronuclear form with four MDSs $M_1, \ldots, M_4$, where $\bar{M}_i$ denotes the inverse of M_i (M_i rotated by 180 degrees). In the macronuclear form of this gene the MDSs are sorted to obtain the consecutive sequence $M_1 M_2 \cdots M_\kappa$.

In general, the structure of MDS M_i for $1 < i < \kappa$, M_1 and M_κ can be depicted as:

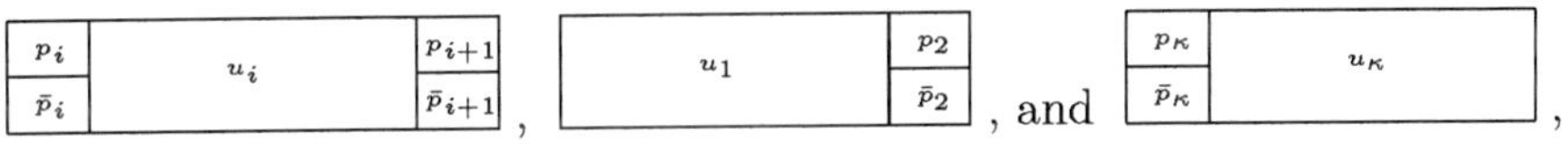

respectively. The symbols p_i and $\bar{p}_i$ for $2 \leq i \leq \kappa$ represent single stranded DNA sequences, and u_i for $2 \leq i \leq \kappa$ represent double stranded DNA sequences. DNA sequence $\bar{p}_i$ is the Watson-Crick complement of DNA sequence p_i. Each double strand $P(i) = \begin{matrix} p_i \\ \bar{p}_i \end{matrix}$ is called a *pointer*. In the macronucleus, the MDSs $M_1, M_2, \ldots, M_\kappa$ are spliced together by 'gluing' each M_j with M_{j+1} on $P(j+1)$ (for $1 \leq j < \kappa$) to obtain the consecutive sequence $M_1 M_2 \cdots M_\kappa$. This gluing is irreversible and thus after gluing M_j with M_{j+1}, the occurrences of $P(j+1)$ in M_j and M_{j+1} are not considered pointers anymore. Consequently, the molecular operations can be seen as operations that remove pointers. We can abstract from the above sequence of MDSs by considering only the pointers in the sequence, we denote i as the pointer $P(i)$:

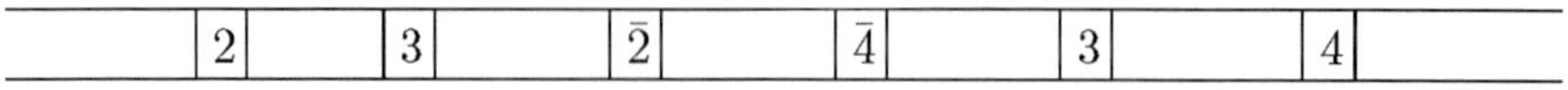

In this way we obtain the string $23\bar{2}\bar{4}34$ as a representation of the micronuclear gene. Such a string is called a *legal string* since it is a string over $\Pi = \{2, 3, \ldots, \kappa\} \cup \{\bar{2}, \bar{3}, \ldots, \bar{\kappa}\}$ and for each $p \in \Pi$ that occurs in u, u contains exactly two occurrences from $\{p, \bar{p}\}$. We will use the convention that $\bar{\bar{p}} = p$ for $p \in \Pi$. We also define $\Delta = \{2, 3, \ldots, \kappa\}$.

Example 1. The gene (in its micronuclear form) that encodes the actin protein in the stichotrich *Sterkiella nova* is described by legal string $3445675678 9\bar{3}\bar{2}289$ (see [10], [11], and [9] for the corresponding sequence of MDSs).

The gene assembly process is accomplished through three molecular operations called loop, hairpin, and double-loop recombination. They are depicted below (in this order).

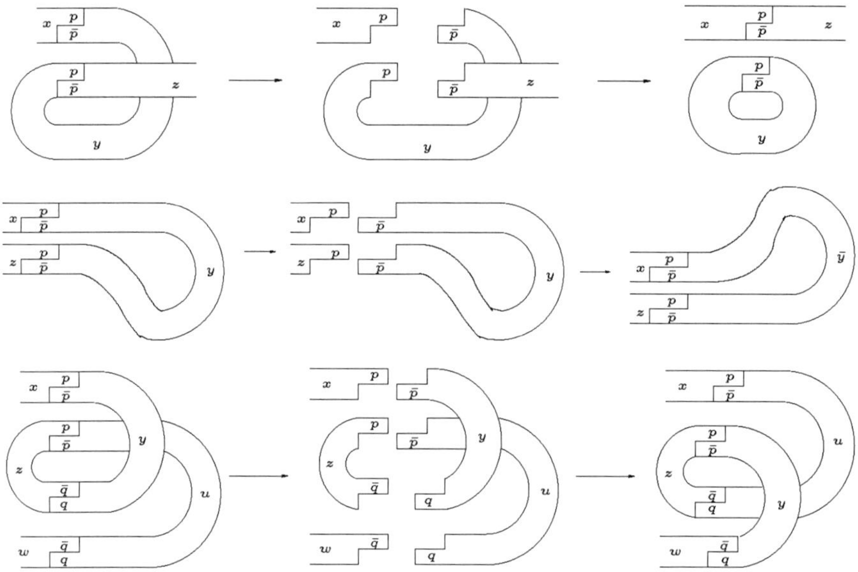

Loop recombination can (only) be applied on a sequence containing a pair of identical pointers without any pointers in between (because a MDS should not be removed from the molecule), hairpin recombination can (only) be applied on a sequence containing a pair of pointers in which one pointer is an inversion of the other, and the double-loop recombination can (only) be applied on a sequence containing two pairs of identical pointers for which the sequence between the first pair of pointers overlaps with the sequence between the second pair of pointers.

The effect of these operations on legal strings is described using the string rewriting rules, called *string pointer rules*. First we define for a string $u = x_1 x_2 \cdots x_n$ with $x_i \in \Pi$, $\bar{u} = \bar{x}_n \bar{x}_{n-1} \cdots \bar{x}_1$. Also, for $p \in \Pi$, we define $\mathbf{p} = p$ if $p \in \Delta$ and $\mathbf{p} = \bar{p}$ if $p \in \bar{\Delta}$, i.e., $\mathbf{p}$ is the 'unbarred' variant of p. We now define the string pointer rules. For all $p, q \in \Pi$ with $\mathbf{p} \neq \mathbf{q}$:

- the *string negative rule* for p is defined by $\mathbf{snr}_p(u_1 pp u_2) = u_1 u_2$,
- the *string positive rule* for p is defined by $\mathbf{spr}_p(u_1 p u_2 \bar{p} u_3) = u_1 \bar{u}_2 u_3$,
- the *string double rule* for p, q is defined by $\mathbf{sdr}_{p,q}(u_1 p u_2 q u_3 p u_4 q u_5) = u_1 u_4 u_3 u_2 u_5$,

where $u_1, u_2, \ldots, u_5$ are arbitrary strings over Π.

Example 2. We consider legal string $u = 344567567 89\bar{3}\bar{2}289$ from Example 1. We cannot apply $\mathbf{spr}_{\bar{8}}$ to u, but we can apply $\mathbf{spr}_3$ to u. We have $\mathbf{spr}_3(u) = \bar{9}\bar{8}\bar{7}\bar{6}\bar{5}\bar{7}\bar{6}\bar{5}\bar{4}\bar{4}\bar{2}289$. Now, we can apply $\mathbf{spr}_{\bar{8}}$ to the obtained legal string.

Note that each of these rules is defined only on legal strings that satisfy the given form. For example, $\mathbf{snr}_2$ is not defined on legal string 2323.

The string negative (positive, double, resp.) rule corresponds to the loop (hairpin, double-loop, resp.) recombination operation. Note that the fact that the molecular operations remove pointers is explicit in the string pointer rules. The model constituting the notions of legal string and the three types of string pointer rules is called string pointer reduction system (**SPRS**, for short) and was introduced in [12]. Also, monograph [9] describes this model in detail.

For a given gene in its micronuclear form, a sequence of these molecular operations is *successful* if it transforms the pattern into its macronuclear form. Note that the macronuclear form of a gene corresponds to the empty string. We now formally define the notion of successfulness. Let u and v be legal strings. A composition φ of reduction rules is called a *reduction of* u, if φ is applicable to (defined on) u. A *successful reduction* φ *of* u is a reduction of u such that $\varphi(u) = \lambda$, where λ denotes the empty string. We then also say that φ is *successful for* u. Thus, e.g., $\mathbf{snr}_4\ \mathbf{spr}_{\bar{3}}\ \mathbf{spr}_2$ is a successful reduction for $23\bar{2}\bar{4}34$. Every legal string has a successful reduction (cf. Theorem 9.1 in [9]).

The *domain* of a string $v \in \Pi^*$ is $\mathrm{dom}(v) = \{\mathbf{p} \mid p \text{ occurs in } v\}$. For a pointer p and a legal string u, if both p and $\bar{p}$ occur in u then we say that both p and $\bar{p}$ are *positive* in u; if on the other hand only p or only $\bar{p}$ occur in u, then both p and $\bar{p}$ are *negative* in u (hence the names string negative rule and string positive rule). So, every pointer occurring in a legal string is either positive or negative in it. We define $\mathrm{dom}(\rho)$ for a reduction rule ρ by $\mathrm{dom}(\mathbf{snr}_p) = \mathrm{dom}(\mathbf{spr}_p) = \{\mathbf{p}\}$ and $\mathrm{dom}(\mathbf{sdr}_{p,q}) = \{\mathbf{p}, \mathbf{q}\}$ for $p, q \in \Pi$. For a composition $\varphi = \rho_n\ \cdots\ \rho_2\ \rho_1$ of reduction rules $\rho_1, \rho_2, \ldots, \rho_n$, we define $\mathrm{dom}(\varphi) = \mathrm{dom}(\rho_1) \cup \mathrm{dom}(\rho_2) \cup \cdots \cup \mathrm{dom}(\rho_n)$. Moreover, we define $\mathrm{snrdom}(\varphi) = \mathrm{dom}(\rho_{i_1}) \cup \cdots \cup \mathrm{dom}(\rho_{i_k})$, where $\rho_{i_1}, \ldots, \rho_{i_k}$ are the **snr** rules of φ.

3 Reduction Graph

A *2-edge coloured graph* is a 6-tuple (V, E_1, E_2, f, s, t), where V is a finite set of *vertices*, $s, t \in V$ are called the *source* and *target*, and $f : V \backslash \{s, t\} \to \Gamma$ is a vertex labelling function. There a two (not necessary disjoint) sets of edges $E_1, E_2 \subseteq \{\{x, y\} \mid x, y \in V, x \neq y\}$. 2-edge coloured graphs G and G' are considered *isomorphic*, denoted $G \approx G'$, when they are equal modulo the identity of the vertices. However, in this case, the labels of the identified vertices must be equal. Also, the source (target, resp.) of G needs to be identified with the source (target, resp.) of G'.

We are now ready to recall the definition of reduction graph. It was introduced in [4], and we restate it here in a less general form. A reduction graph is a 2-edge coloured graph where the two types of edges are called *reality edges* and *desire edges*. Moreover, all vertices, except for two distinct vertices s and t, are labelled by an element from Δ. As an example, consider the physical representation of legal string $u = 2\bar{7}47353\bar{4}2656$ given below:

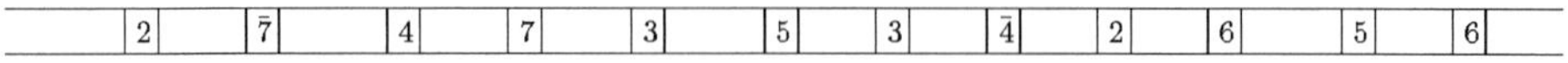

We will use this legal string as our running example. The reduction graph for this u is given below.

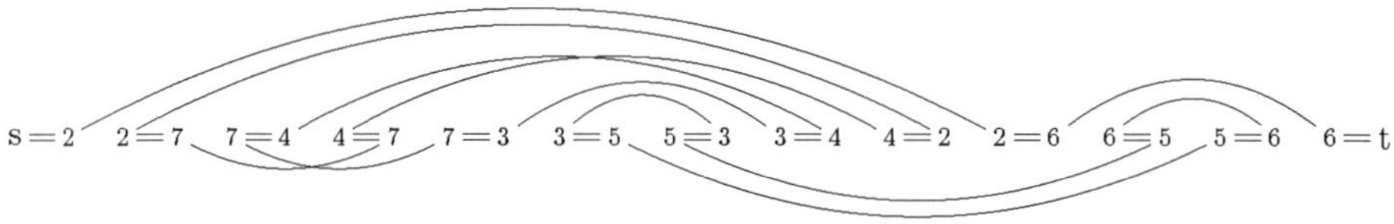

A reduction graph for a legal string u is defined in such a way that (1) each (occurrence of a) pointer of u appears twice (in unbarred form) as a vertex in the graph to represent both sides of the pointer in the physical representation of u, (2) the reality edges (depicted as 'double edges' to distinguish them from the desire edges) represent the segments between the pointers, (3) the desire edges represent which segments should be glued to each other when operations are applied on the corresponding pointers. To enforce this last requirement, positive pointers are connected by crossing desire edges (cf. pointer 7 in the figure above), while negative pointers are connected by parallel desire edges. The notion is similar to the breakpoint graph (or reality-and-desire diagram) known from the theory of sorting by reversal, see e.g. [1] and [2].

Definition 3. Let $u = p_1 p_2 \cdots p_n$ with $p_1, \ldots, p_n \in \Pi$ be a legal string. The *reduction graph of* u, denoted by $\mathcal{R}_u$, is a 2-edge coloured graph (V, E_1, E_2, f, s, t), where

$$V = \{I_1, I_2, \ldots, I_n\} \cup \{I'_1, I'_2, \ldots, I'_n\} \cup \{s, t\},$$

$$E_1 = \{e_0, e_1, \ldots, e_n\} \text{ with } e_i = \{I'_i, I_{i+1}\} \text{ for } 1 < i < n, e_0 = \{s, I_1\}, e_n = \{I'_n, t\},$$

$$E_2 = \{\{I'_i, I_j\}, \{I_i, I'_j\} \mid i, j \in \{1, 2, \ldots, n\} \text{ with } i \neq j \text{ and } p_i = p_j\} \cup \\ \{\{I_i, I_j\}, \{I'_i, I'_j\} \mid i, j \in \{1, 2, \ldots, n\} \text{ and } p_i = \bar{p}_j\}, \text{ and}$$

$$f(I_i) = f(I'_i) = \mathbf{p}_i \text{ for } 1 \leq i \leq n.$$

The edges of E_1 are the reality edges, and the edges of E_2 are the desire edges. In depictions of reduction graphs, we will represent the vertices (except

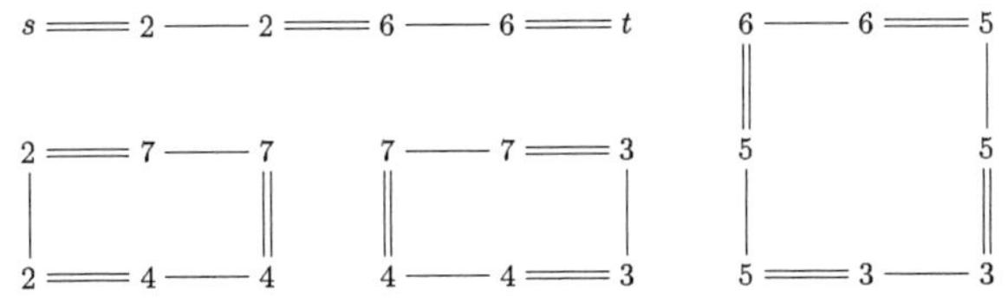

Fig. 1. The reduction graph $\mathcal{R}_u$ of u from the running example.

for s and t) by their labels, because the exact identities of the vertices is not essential for the problems considered in this paper.

In the running example, the reduction graph $\mathcal{R}_u$ of u is again depicted in Figure 1 — we have only rearranged the vertices.

Each reduction graph has a connected component with a linear structure containing both the source and the target vertex [4]. This connected component is called the *linear component* of the reduction graph. The other connected components are called *cyclic components* because of their structure. In sequel of this paper we discuss the uses of the reduction graph for **SPRS**.

4 Cyclic Components

Since the reality edges represent segments of DNA and the desire edges represent how the segments are glued after gene assembly is performed, for a given legal string u (which represents the micronuclear form of a gene) the reduction graph of u represents the macronuclear form of that gene. Therefore the cyclic components of a reduction graph represent circular molecules. If we now again consider the recombination operations in Section 2, it is clear that each such molecule is obtained by loop recombination. Notice that, since the reduction graph is fixed for a given u, the end product after gene assembly is fixed. There can be many different sequences of operations that obtain the fixed end product, however we have shown now that the *number* of loop recombination operations (string negative rules in the model) in each such sequence is the same.

Theorem 4. *Let N be the number of cyclic components in the reduction graph of legal string u. Then every successful reduction of u has exactly N string negative rules.*

Example 1. Since $\mathcal{R}_u$ in Figure 1 has three cyclic components, by Theorem 4, every successful reduction φ of u has exactly three string negative rules. For example $\varphi = \mathbf{snr}_2\ \mathbf{snr}_{\bar{4}}\ \mathbf{spr}_{\bar{7}}\ \mathbf{snr}_6\ \mathbf{sdr}_{3,5}$ is a successful reduction of u. Indeed, φ has exactly three string negative rules. Alternatively, $\mathbf{snr}_6\ \mathbf{snr}_3\ \mathbf{snr}_7\ \mathbf{spr}_2\ \mathbf{spr}_{\bar{5}}\ \mathbf{spr}_4$ is also a successful reduction of u, with a different number of (**spr** and **sdr**) operations.

It turns out that the reduction graph also allows for determining *on which pointers* the string negative rules can be applied. To show this we need some definitions. If it is clear from the context which legal string u is meant, we will denote the set of connected components of the reduction graph of u by ζ. A *multigraph* is a (undirected) graph $G = (V, E, \epsilon)$, where parallel edges are possible. Therefore, E is a finite set of edges and $\epsilon : E \to \{\{x, y\} \mid x, y \in V\}$ is the *endpoint mapping*. We allow $x = y$, and therefore edges can be of the form $\{x, x\} = \{x\}$ — an edge of this form should be seen as a 'loop' for x.

We now define a graph given a reduction graph that represents how the labels of that reduction graph are distributed among its connected components.

Definition 5. Let u be a legal string. The *pointer-component graph of* u *(or of* $\mathcal{R}_u$*)*, denoted by $\mathcal{PC}_u$, is a multigraph (ζ, E, ϵ), where $E = \mathrm{dom}(u)$ and ϵ is, for $e \in E$, defined by $\epsilon(e) = \{C \in \zeta \mid C \text{ contains vertices labelled by } e\}$.

The pointer-component graph of the reduction graph of Figure 1 is given below. We have $\zeta = \{C_1, C_2, C_3, R\}$ where R is the linear component and the other elements are cyclic components of $\mathcal{R}_u$.

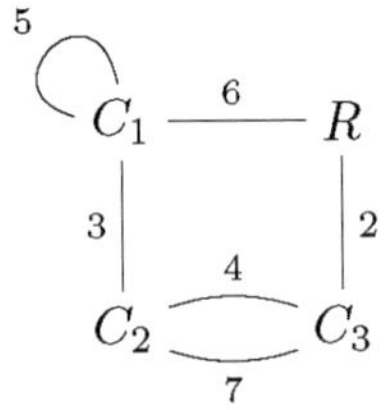

Let us denote $\mathcal{PC}_u|_D$ as the graph obtained from $\mathcal{PC}_u$ by removing the edges outside D.

Theorem 6. *Let u be a legal string, and let $D \subseteq \mathrm{dom}(u)$. There is a successful reduction φ of u with* $\mathrm{snrdom}(\varphi) = D$ *iff $\mathcal{PC}_u|_D$ is a tree.*

In our running example, we see that $D = \{2, 3, 6\}$ induces a (spanning) tree of $\mathcal{PC}_u$. Therefore there is a successful reduction φ of u with $\mathrm{snrdom}(\varphi) = D$. Indeed, we have $\varphi' = \mathbf{sdr}_{\bar{4},5}\ \mathbf{spr}_{\bar{7}}(u) = 226336$. It is clear that we can extend φ' to a successful reduction which applies string negative rules on 2, 3, and 6. Notice that here $\mathbf{snr}_3$ must be applied *before* $\mathbf{snr}_6$. In fact in [6] it is shown that the possible orders in which the string negative rules can be applied is also deducible from $\mathcal{PC}_u$ by considering rooted trees.

5 The Structure of Macronuclear Genes with Spliced Molecules

Since the reduction graph is by definition a representation of the macronuclear form of a gene, it gives rise to the question: Which graphs are (isomorphic to) reduction graphs? This may help to identify the possible structures of the macronuclear form of genes. We answer the question here.

Certainly, a graph G isomorphic to a reduction graph must be a 2-edge coloured graph (V, E_1, E_2, f, s, t) such that for each p in the range of f, $p \in \Delta$

and there must be exactly 4 vertices labelled by p. Each vertex must be connected to exactly one (reality) edge from E_1, and each vertex, except s and t, must be connected to exactly one (desire) edge from E_2. Finally, edges from E_2 must connect vertices with a common label. Let us call the set of graphs with these properties $\mathcal{G}$. It turns out that there are graphs in $\mathcal{G}$ that are not (isomorphic to) reduction graphs.

To obtain a characterization we need one more property of reduction graphs: the possibility to linearly order the vertices as done in a figure in Section 3. To make this linear order of vertices explicit, we introduce *merge edges* to the reduction graph as done below.

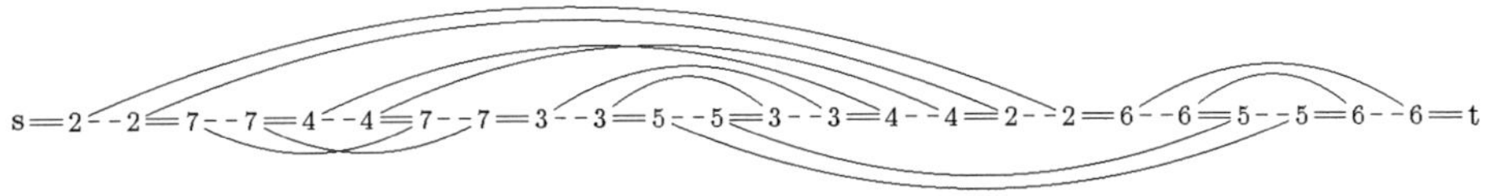

Remark. In the breakpoint graph the reality edges are actually the merge edges in the reduction graph. Thus perhaps it would have been more appropriate to call merge edges reality edges, and reality edges e.g. "segment edges" (since they represent segments of DNA) in the reduction graph. The notion of segment edges is not introduced in breakpoint graphs but it is implicitly present since the graph is drawn as a circular diagram: the reality and desire diagram.

Now, when is a set of edges M for $G \in \mathcal{G}$ a set of merge edges? Like desire edges they have the properties that (1) the edges connect vertices with a common label and (2) each vertex except s and t is connected to exactly one merge edge. Moreover, M and the set E_2 are disjoint — no desire edge is parallel to a merge edge. Finally, the reality edges and merge edges must allow for a path from s to t passing each vertex once. This last requirement is equivalent to the fact that the reality and merge edges induce a connected graph.

If it is possible to add a set of merge edges to the graph, then it is not difficult to see that the graph is isomorphic to a reduction graph $\mathcal{R}_u$. Indeed, we can identify such a u for this reduction graph by simply considering the alternating path from s to t over the reality and merge edges. The orientation (positiveness or negativeness) of each pointer is determined by the crossing or non crossing of the desire edges (exactly as in the definition of reduction graph).

It is shown in [6] that, surprisingly, $G \in \mathcal{G}$ has a set of merge edges precisely when the pointer-component graph $\mathcal{PC}_G$ is a connected graph (the definition of $\mathcal{PC}_G$ is a trivial generalization of Definition 5). In other words, G is (isomorphic to) a reduction graph iff $G \in \mathcal{G}$ and $\mathcal{PC}_G$ is a connected graph.

Let us consider now legal strings $u = pq\bar{p}q$ and $v = pqpq$. It turns out that they have the same reduction graph (up to isomorphism): $\mathcal{R}_u \approx \mathcal{R}_v$. The reason for this is that a reduction graph may have more than one set of merge edges — each one corresponding to a different legal string, cf. Section 5. Thus there can be many micronuclear (forms of) genes (i.e., legal strings) obtaining the same macronuclear structure (i.e., reduction graph). In [6] it is shown how for a given legal string u we can obtain precisely the set of all legal strings having the same reduction graph (up to isomorphism). In fact, it turns out that this set is exactly the set of all legal strings obtained by applying compositions of the following string rewriting rules.

For all $p, q \in \Pi$ with $\mathbf{p} \neq \mathbf{q}$ we define

- the *dual string positive rule* for p is defined by $\mathbf{dspr}_p(u_1 p u_2 p u_3) = u_1 p \bar{u}_2 p u_3$,
- the *dual string double rule* for p, q is defined by $\mathbf{dsdr}_{p,q}(u_1 p u_2 q u_3 \bar{p} u_4 \bar{q} u_5) = u_1 p u_4 q u_3 \bar{p} u_2 \bar{q} u_5$,

where $u_1, u_2, \ldots, u_5$ are arbitrary (possibly empty) strings over Π.

Example 7. Let us take $u = pq\bar{p}q$ and $v = pqpq$ given earlier. Since $\mathbf{dspr}_q(u) = v$, both legal strings indeed have a common reduction graph.

6 Intermediate Gene Patterns

We have shown that the reduction graph is a representation of the end product of gene assembly for a given gene in its micronuclear form (given by a legal string u). Here we show that we can generalize the notion of reduction graph to allow for representations of every intermediate product. To do this, we fix a subset D of $\mathrm{dom}(u)$ as the set of pointers which have not yet been used in the recombination operations, while the pointers outside D (thus in $\mathrm{dom}(u) \backslash D$) have already been used. We can then define the reduction graph $\mathcal{R}_{u,D}$ of u w.r.t. D as illustrated below in our running example with $D = \{2, 4\}$.

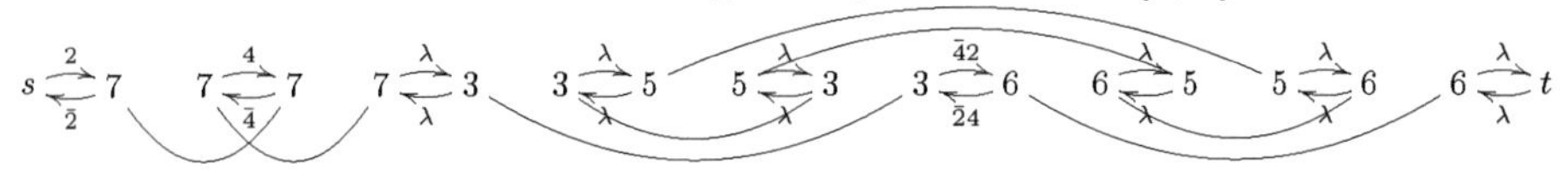

We simply ignore the pointers in D — they are put as strings on the reality edges which are now directed edges. The reduction graph of u w.r.t. $\varnothing$ is equivalent to the reduction graph of u in Definition 3. The next result is a reformulation of Theorem 18 of [4]. We will denote the legal string obtained from a legal string u by removing the pointers from $D \subseteq \mathrm{dom}(u)$ and its barred variants by $\mathrm{rem}_D(u)$. Thus, e.g., $\mathrm{rem}_{\{2,4\}}(\bar{2}34\bar{3}\bar{2}4) = 3\bar{3}$. We define $\mathrm{red}(u, D)$ as the label of the alternating path from s to t. Thus in our example $\mathrm{red}(u, D) = 2\bar{4}\bar{4}2$.

Theorem 8. *Let u be a legal string, let φ be a composition of reduction rules with $\mathrm{dom}(\varphi) \subseteq \mathrm{dom}(u)$, and let $D = \mathrm{dom}(u) \backslash \mathrm{dom}(\varphi)$. Then φ is applicable to u iff φ is applicable to $\mathrm{rem}_D(u)$ and $\mathrm{red}(u, D)$ is legal with domain D. Moreover, if this is the case, then $\varphi(u) = \mathrm{red}(u, D)$.*

As a consequence, reductions φ_1 and φ_2 with the same domain have the same effect: $\varphi_1(u) = \varphi_2(u)$ for all legal strings u. This puts Theorem 4.8 of [13] into a new perspective: the fact that a reordering of operations yields the same result is independent of the parallelism of the operations. Also note that in general there are $D \subseteq \mathrm{dom}(u)$ for which there is no reduction φ of u with $D = \mathrm{dom}(\varphi(u))$. In our example $\mathrm{red}(u, D)$ is legal with domain D and we have e.g. $(\mathbf{snr}_6\ \mathbf{sdr}_{3,5}\ \mathbf{spr}_{\bar{7}})(u) = 2\bar{4}\bar{4}2 = \mathrm{red}(u, D)$.

The results of the previous sections can be carried over to this general case. E.g., the number of string negative rules from u to $\varphi(u)$ is fixed and is equal to the number of cyclic components of $\mathcal{R}_{u,D}$.

7 Discussion

The concept of breakpoint graph from the theory of sorting by reversal has many uses for gene assembly in ciliates. While a legal string is a representation of the micronuclear form of a gene (the begin product of gene assembly), the reduction graph of that legal string is a representation of the macronuclear form of that same gene (the end product of gene assembly). The possible forms of reduction graphs are characterized, as well as the sets of legal strings that obtain a common reduction graph. Also, loop recombination is well explained by the reduction graph, just as the reversal operation (similar to hairpin recombination) is explained by the breakpoint graph. It remains an open problem to characterize hairpin and double-loop recombination. Similarly, sorting by block-interchange (similar to double-loop) is very much open.

References

1. Setubal, J., Meidanis, J.: Introduction to Computional Molecular Biology. PWS Publishing Company (1997)
2. Pevzner, P.: Computational Molecular Biology: An Algorithmic Approach. MIT Press (2000)
3. Bergeron, A., Mixtacki, J., Stoye, J.: On sorting by translocations. In Miyano, S., et al., eds.: RECOMB. Volume 3500 of LNCS., Springer (2005) 615–629
4. Brijder, R., Hoogeboom, H., Rozenberg, G.: Reducibility of gene patterns in ciliates using the breakpoint graph. Theor. Comput. Sci. **356** (2006) 26–45
5. Brijder, R., Hoogeboom, H., Muskulus, M.: Strategies of loop recombination in ciliates. LIACS Technical Report 2006-01, [arXiv:cs.LO/0601135] (2006)
6. Brijder, R., Hoogeboom, H.: The fibers and range of reduction graphs in ciliates. LIACS Technical Report 2007-01, [arXiv:cs.LO/0702041] (2007)
7. Brijder, R., Hoogeboom, H., Rozenberg, G.: The breakpoint graph in ciliates. In Berthold, M., et al., eds.: CompLife '05. Volume 3695 of LNCS., Springer (2005) 128–139
8. Brijder, R., Hoogeboom, H., Muskulus, M.: Applicability of loop recombination in ciliates using the breakpoint graph. In Berthold, M., et al., eds.: CompLife '06. Volume 4216 of LNCS., Springer (2006) 97–106
9. Ehrenfeucht, A., Harju, T., Petre, I., Prescott, D., Rozenberg, G.: Computation in Living Cells – Gene Assembly in Ciliates. Springer Verlag (2004)
10. Prescott, D., DuBois, M.: Internal eliminated segments (IESs) of oxytrichidae. J. Euk. Microbiol. **43** (1996) 432–441
11. Cavalcanti, A., Clarke, T., Landweber, L.: MDS_IES_DB: a database of macronuclear and micronuclear genes in spirotrichous ciliates. Nucleic Acids Res. **33** (2005) D396–D398
12. Ehrenfeucht, A., Harju, T., Petre, I., Prescott, D., Rozenberg, G.: Formal systems for gene assembly in ciliates. Theor. Comput. Sci. **292** (2003) 199–219
13. Harju, T., Li, C., Petre, I., Rozenberg, G.: Parallelism in gene assembly. Natural Computing **5** (2006) 203–223

Bayesian median regression for temporal gene expression data

Keming Yu[1*], Veronica Vinciotti[2*], Xiaohui Liu[2] and Peter A.C. 't Hoen[3]
1. Department of Mathematics, Brunel University, Uxbridge UB8 3PH, UK
2. Department of Information Systems and Computing, Brunel University, Uxbridge UB8 3PH, UK
3.Center for Human and Clinical Genetics, Leiden University Medical Center, Postzone S4-0P, P.O. Box 9600, 2300 RC Leiden, Netherlands

Abstract. Most of the existing methods for the identification of biologically interesting genes in a temporal expression profiling dataset do not fully exploit the temporal ordering in the dataset and are based on normality assumptions for the gene expression.
In this paper, we introduce a Bayesian median regression model to detect genes whose temporal profile is significantly different across a number of biological conditions. The regression model is defined by a polynomial function where both time and condition effects as well as interactions between the two are included. MCMC-based inference returns the posterior distribution of the polynomial coefficients. From this a simple Bayes factor test is proposed to test for significance.
The estimation of the median rather than the mean, and within a Bayesian framework, increases the robustness of the method compared to a Hotelling T^2-test previously suggested. This is shown on simulated data and on muscular dystrophy gene expression data.

1 Background

In a typical time course microarray study, a number of microarray experiments are carried out at biologically interesting time points and across different biological conditions (e.g. normal versus disease, different forms of a disease). It is a frequent and challenging goal to try to identify which of these genes exhibit an interesting temporal behaviour, for example whether and when a gene becomes up- or down-regulated and, more importantly, whether its behaviour is significantly different across the biological conditions of interest.

Various methods have been proposed in the literature to detect differentially expressed genes from time course microarray experiments. However, only few studies have looked at the problem of detecting genes with significantly different temporal profiles across a number of biological conditions (e.g. [1] and reference therein). Furthermore, most of these studies are based on normality assumptions of the gene expression across different arrays, which is often violated given the

* These authors equally contributed to this work.

CP940, *CompLife 2007, 3rd International Symposium on Computational Life Science*
edited by A. Siebes, M. Berthold, R. Glen, and A. Feelders

small number of observations per gene typically available in a microarray study. [2] is one of the few studies where this assumption is not made and inference is made on the quantiles of the gene expression distribution.

In [1], we propose a statistical test to detect genes with temporal behaviour significantly different across a number of experimental groups. We account for the temporal order in the data by fitting polynomial models to single gene profiles for each of the groups. We derive a Hotelling T^2-statistic to detect the genes whose fitted profiles are significantly different from each other, based on a normal approximation of the gene expression.

In this paper, we extend this approach in different directions. First of all, we model the *median* of the gene expression, rather than the mean, in order to improve the robustness of the method against outliers and abnormalities, which are quite common in gene expression data. A regression model is used to describe the median: the model is described by a polynomial function with respect to time, where the polynomial coefficients depend on the biological condition. In this way, the model accounts for time and condition effects as well as possible interactions between the two. Secondly, we make inference on the model parameters using a *Bayesian* approach. Bayesian inference provides a simple and easy method for obtaining the full posterior distributions (not only single values) of the regression coefficients. Bayesian approaches have been used before to test for differential gene expression (e.g.[3] and references therein) though none of these studies has considered time-course microarray data. In this paper, we propose a Bayes factor test to detect the genes for which the temporal profiles are significantly different across the different biological conditions. Finally, the proposed approach does not need any normality assumption for the error term, which can take any form. As such it is more suited to this application, as the distribution of gene expression is generally non-Gaussian.

In section 2, we describe the regression model, the Bayesian inference of the regression coefficients and the Bayesian statistical test for the detection of the parameters significantly different from zero. In section 3, we compare this approach with the Hotelling T^2-test in [1]. We carry out the comparison both on simulated data and on time-course microarray data from four different forms of muscular dystrophy. Finally, in section 4 we draw some conclusions and highlight the potentials of this new methodology.

2 Methods

2.1 Linear regression model of temporal profiles

In this paper, we introduce a linear regression model to detect the genes for which the temporal profiles across different conditions are significantly different, that is the genes that show a significant effect between conditions. The temporal aspect of the data is captured using a polynomial of degree p. An interaction term between conditions and time is also included in the model, as we are interested in genes that show a different temporal behaviour across conditions and not simply in genes that are constantly over-expressed in one condition and under-expressed in the other.

Let y_{tc} denote the gene expression at time t under the biological condition c and for a particular gene. For simplicity, we avoid referring to the specific gene in the notation. Normally, y_{tc} is in the form of a log-ratio with respect to a reference channel. In section 2.4, we will extend the model to the case of any microarray experimental design. Let μ_{ct} denote the true gene expression at time t with respect to the initial time point and under condition c. The vector $\mu_c = (\mu_{c2}, \cdots, \mu_{cT})$ for all time points t is the true temporal profile for a specific gene under the biological condition c. This is what we wish to estimate from the available data.

We consider the following standard linear model

$$y_{ct} = \mu(\boldsymbol{x}_{ct}) + \epsilon_{ct}, \tag{1}$$

where $\mu(\boldsymbol{x}_{ct})$ may be thought of as the conditional mean of y_{ct} given the vector of regressors $\boldsymbol{x}_{ct}$, and ϵ_{ct} is the error term with mean zero and constant variance. Then we model the gene expression at time t for condition c, where $c = \{1, \cdots, C\}$, as a polynomial of degree p

$$\mu_{ct} = \mu(\boldsymbol{x}_{ct}) = \beta_{0c} + \beta_{1c}\, t + \cdots + \beta_{cp}\, t^p. \tag{2}$$

Note that the coefficients of the polynomial depend on the condition c, that is a different polynomial is allowed for each condition. Note also that a different model is fit for each gene.

Given the presence of abnormalities and possible heteroscedastity of gene expression, which one can see for example from Figure 1, we propose a Bayesian median regression model [4, 5] for the estimation of the parameters in (2), to replace the normal distribution based estimation more commonly used (e.g. [1]). The use of the median rather than the mean is also expected to be more robust to small sample size problems, which is the case for gene expression data.

The next section describes the inference approach for the estimation of the vector of regression parameters β.

2.2 Bayesian median inference with MCMC algorithm

In strictly linear models, a simple approach to estimating the conditional quantiles is suggested in [6]. Consider the standard linear model of equation (1). Here, it is not necessary to specify the distribution of the error term as it is allowed to take any form. Typically,

$$\mu(\boldsymbol{x}) = \boldsymbol{x}'\boldsymbol{\beta}$$

for a vector of coefficients $\boldsymbol{\beta}$. The τth $(0 < \tau < 1)$ quantile of ϵ is the value, q_τ, for which $P(\epsilon < q_\tau) = \tau$. The τth conditional quantile of y_t given $\boldsymbol{x}$ is then simply

$$q_\tau(y_t|\boldsymbol{x}) = \boldsymbol{x}'\boldsymbol{\beta}(\tau),$$

where $\boldsymbol{\beta}(\tau)$ is a vector of coefficients dependent on τ.

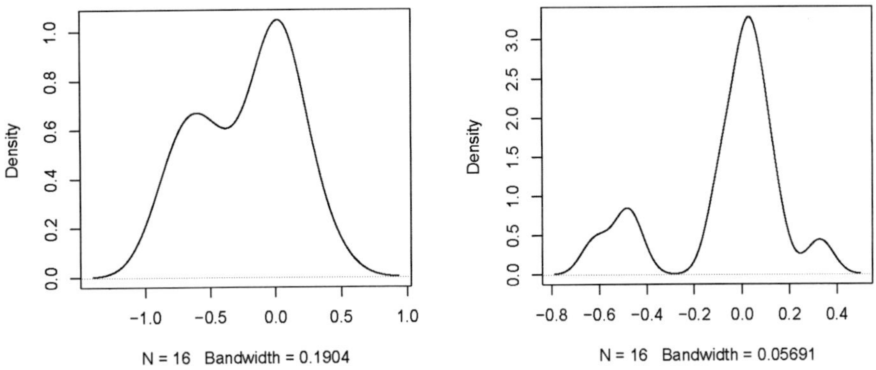

Fig. 1. Kernel estimation of the gene expressions under two different conditions for a representative gene.

The τth *regression quantile* $(0 < \tau < 1)$ is defined as any solution, $\hat{\boldsymbol{\beta}}(\tau)$, to the *quantile regression* minimisation problem

$$\min_{\boldsymbol{\beta}} \sum_t \rho_\tau(y_t - \boldsymbol{x}'\boldsymbol{\beta}),$$

where the loss function

$$\rho_\tau(u) = u(\tau - I(u < 0)).$$

The median corresponds to $\tau = \frac{1}{2}$.

Given the observations, $\boldsymbol{y} = (y_1, \ldots, y_n)$, the posterior distribution of $\boldsymbol{\beta}$, $\pi(\boldsymbol{\beta}|\boldsymbol{y})$ is given by

$$\pi(\boldsymbol{\beta}|\boldsymbol{y}) \propto L(\boldsymbol{y}|\boldsymbol{\beta})\,\pi(\boldsymbol{\beta}), \tag{3}$$

where $\pi(\boldsymbol{\beta})$ is the prior distribution of $\boldsymbol{\beta}$ and $L(\boldsymbol{y}|\boldsymbol{\beta})$ is the likelihood function written as

$$L(\boldsymbol{y}|\boldsymbol{\beta}) = \tau^n\,(1-\tau)^n\,\exp\left\{-\sum_i \rho_\tau(y_i - \boldsymbol{x}_i'\boldsymbol{\beta})\right\}, \tag{4}$$

which can be regarded based on an asymmetric Laplace density with a location parameter $\mu_i = \boldsymbol{x}_i'\boldsymbol{\beta}$.

In theory we could use any prior in (3), but in the absence of any realistic information one could use improper uniform prior distributions for all the components of $\boldsymbol{\beta}$. This choice is appealing as the resulting joint posterior distribution is proportional to the likelihood surface. The posterior distribution of the data is computed using MCMC methods [7].

2.3 Bayesian hypothesis test

Once the parameters of the polynomial are estimated, the aim is to detect the genes whose parameters of the temporal profiles are significantly different across the biological conditions. Bayes factors are the dominant method of Bayesian model testing. They are the Bayesian analogues of likelihood ratio tests and they are regarded as a simple and robust method for testing. The basic intuition is that prior and posterior information are combined in a ratio that provides evidence in favour of one model specification versus another.

Suppose the observed gene data $\boldsymbol{y}$ are generated under a model M_i, one of a set $\mathcal{M}$ of computing models. Each model specifies the distribution of $\boldsymbol{y}$, $f(Y|\boldsymbol{\beta}_i, M_i)$ apart from an unknown parameter vector $\boldsymbol{\beta}_i$. Suppose we have some preliminary knowledge about condition probabilities $\pi(M_i)$. The interest lies in obtaining the posterior probabilities $\pi(M_i|\boldsymbol{y})$ for every model M_i. A choice between two models (say models M_1 and M_2) is based on the Bayes factor

$$B_{12} = \frac{\pi(M_1|\boldsymbol{y})/\pi(M_2|\boldsymbol{y})}{\pi(M_1)/\pi(M_2)},$$

i.e., the ratio of posterior odds of the condition M_1 to prior odds of M_1. The Bayes factor is often thought of as the weight of evidence in favour of model M_1 (see [8] for a review on Bayes factor testing). [9] recommended a scale of evidence for interpreting Bayes factors. In Table 1 we consider $-\log(B_{12})$ and we use a modification of Jeffreys' scale taken from [10] in terms of the negative log-likelihood $-\log(\pi(M_i|\boldsymbol{y})$, which for our problem is given by equation (4).

Table 1. Jeffreys' scale of evidence for Bayes factor

negative-log Bayes factor	Interpretation
$2.3 < -\log(B_{12})$	Strong evidence for M_2
$1.1 <= -\log(B_{12}) <= 2.3$	Moderate evidence for M_2
$0 < -\log(B_{12}) <= 1.1$	Weak evidence for M_2
$-1.1 < -\log(B_{12}) <= 0$	Weak evidence for M_1
$-2.3 < -\log(B_{12}) <= -1.1$	Moderate evidence for M_1
$-\log(B_{12}) <= -2.3$	Strong evidence for M_1

In the context of this paper, we will compare a model where the same parameters of the polynomial are considered for all the biological conditions with a model where these are different. We will be interested in the genes which show significance evidence (based on the scale in Table 1) for the model with different polynomial coefficients, i.e. the genes for which the temporal profiles between at least two biological conditions are significantly different.

2.4 Including a design matrix in the model

The model we have described so far assumes that the data y_{ct} measures the expression of a gene at time t with respect to a reference signal. In general, other designs than the reference design, and normally more efficient than these, can be used [11]. In this case, the data will be in the form of $y_{ct_it_j}$, that is the log-ratio between time t_j and time t_k hybridised on the same array under the biological condition c. In this case, the same methodology described so far can be applied, with the addition of a design matrix X. This matrix defines the relationship between the values observed in the experiment and the parameters of interest, that is the gene expression at time t with respect to the first time point.

The general model will then have the compact matrix form

$$y = X\mu + \epsilon, \tag{5}$$

where X is the design matrix, and $\mu = (\mu_1, .., \mu_C)$, with $\mu_i = (\mu_{i2}, ..., \mu_{iT})$, the parameters of interests for the C conditions and T time points. These are described using the regression model (2).

3 Results

3.1 Simulation study

In [1], we have shown how the Hotelling T^2 temporal test outperforms a non-temporal one. Like the temporal Hotelling T^2-test, our test is of course a temporal one. We have performed a simulation study where we compare the Bayesian median regression test and temporal Hotelling T^2-test.

We have simulated gene expressions for 2 conditions and 9 time points, over 72 arrays and 1000 genes. That is, there are four replicates for each time series. The sample size for this simulation was chosen to reflect the real data in Section 3.2. We have generated the data so that 750 of these genes are not differentially expressed between the two conditions. The remaining genes are assumed to have different temporal profiles between the two conditions, with the temporal profiles linearly increasing in one condition and decreasing in the other. For the error distribution, we considered both the case of a normal distribution $\mathcal{N}(0, \sigma^2)$ and the more realistic case of a mixture of normals distribution $\frac{2}{3}\mathcal{N}(-\mu - 0.1, \sigma^2) + \frac{1}{3}\mathcal{N}(\mu, \sigma^2)$. For all genes, the noise σ was varied uniformly between 0.03 and 1.2, the error mean μ for the mixture of normals between 0.1 and 0.7, and the slope of the linear profile for the differentially expressed genes between 0.005 and 0.2 in absolute value. These values were chosen to reflect the ones found in our real dataset. In the second simulation, we have considered the case of extremely small sample size data with the aim of testing the robustness of the proposed method in these situations. Here we simulated only two time series for each condition, for a total of 36 arrays.

Figure 2 plots the ROC curves for the Hotelling T^2-test and Bayesian median regression test for the simulations described above. For the Bayesian median regression test, 10000 iterations of the MCMC inference were considered which proved sufficient to obtain convergence. A quadratic polynomial for the temporal profiles is used for both tests (that is $p = 2$ in equation 2). Given the different frameworks for the two tests – frequentist and Bayesian, respectively – we have computed the ROC curve as follows: for a fixed threshold for false positive rate, we derive the p-value cutoff for the Hotelling test and the Bayes factor cutoff for the Bayesian test such that the desired false positive rate is reached. This is defined by the proportion of genes falsely predicted as differentially expressed amongst all the genes predicted as differentially expressed. For these cutoffs, the true positive rate is computed, that is the proportion of genes correctly predicted as differentially expressed. Varying the threshold of false positive rate between 0.005 and 0.7 (excluding the extreme values simply for better visualization of the comparison) leads to the plots in Figure 2.

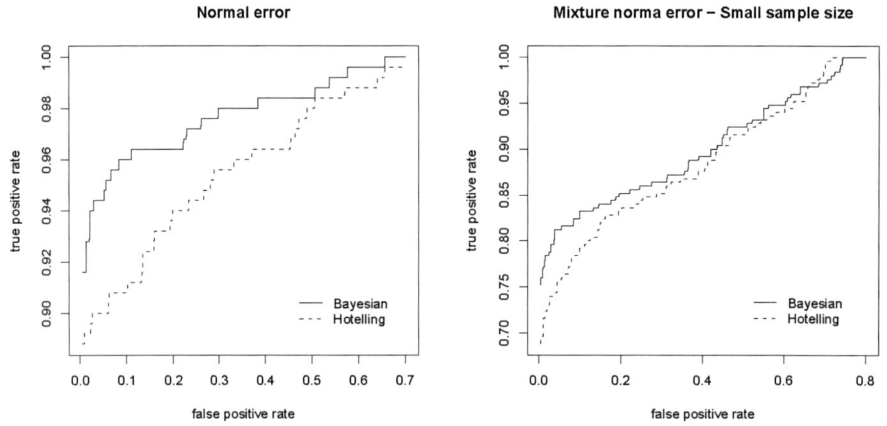

Fig. 2. Comparison of ROC curves for Bayesian (solid line) and Hotelling (dotted line) tests on simulated data: 1000 genes, of which 750 differentially expressed; 2 conditions; 72 arrays and normal error (left); 36 arrays normal and mixture of normals error (right).

Both simulations show an improvement of the proposed method over the previous one, with slightly higher true positive rates for most thresholds of false positive rates. This is promising, especially considering the extra advantages that the new methodology can offer: it is expected to be more robust to noise and small sample size as it estimates the median rather than the mean, it returns the full posterior distribution of the parameters, rather than simply an estimate for the mean, it is not based on any normality assumption and as such it is expected to perform better on a wider number of applications.

In the next section, we will apply the methodology on a real gene expression dataset. In this case, we will use the scale of Table 1 to select the differentially expressed genes. If we were to use this on the simulated data, for example on the second simulation with mixture of normals error and small sample size, we would find 172 genes with strong evidence for differential expression, all correctly predicted. Going down the scale, a further 18 genes would be found with moderate evidence, with one of these being a false positive, and a further 54 genes with weak evidence, with 35 false positives amongst these. So it seems the scale is a good indicator of the strength of evidence in the prediction.

3.2 Analysis of muscular dystrophy dataset

We have applied the Bayesian median regression on microarray data from wild-type mice and three mouse strains with different forms of muscular dystrophy, to identify genes with differences in temporal expression profiles between the strains. The four mouse strains were profiled at different ages: dystrophin-, beta-sarcoglycan and gamma-sarcoglycan deficient mice, and wild-type mice. The first three are animal models for different muscular dystrophies. The data contains 7144 genes, whose expression is measured over 9 time points. For each gene, we have available 16 time series profiles, four for each class. The data and biological problem are further described in [1].

We have calculated Bayes factors for each gene under the Bayesian median regression methodology described in this paper. For this application, we assume the quadratic term of the polynomial to be the same for the four conditions (that is $\beta_{c2} = \beta_2$ for $c = 1, \cdots, 4$ in equation 2). This was done with the aim of reducing the overall number of parameters and as such increasing the robustness and convergence of the MCMC method. Whereas more work is ongoing to improve the MCMC convergence, for this application we did not find this assumption to be particularly restrictive, as the linear terms and interactions between time and conditions managed to capture most of the differences between the temporal profiles across the conditions.

Figure 3 shows the distribution of the Bayes factors for all the genes. As expected, only a small number of genes fall on the right tail of the distribution: these are the differentially expressed genes. Shaded in grey are the Bayes factors for the genes whose p-value from the Hotelling T^2-test is less than 0.05. The two methods show a high agreement, although the Hotelling test seems to find many genes with weak and moderate evidence given by the Bayes factor. The average of the Bayes factors for the genes selected by the Hotelling test was 1.1. The new methodology returned a smaller number of differentially expressed genes. Using the reference scale in Table 1, 96 genes showed strong evidence of differential excpression and a further 221 genes showed moderated evidence. With the Hotelling T^2-test, we had identified 810 genes with multiple testing-corrected p-value less than 0.05. Of the 96 genes found by the Bayesian test with strong evidence, 80 were also found by the Hotelling test.

Our hypothesis is that the Bayesian method returns a less noisy and more robust set of genes with higher relevance to the biological problem under study. To demonstrate this, we evaluated the relevance of the selected genes to the muscular dystrophy phenotype with a literature-based approach and a GeneOntology-

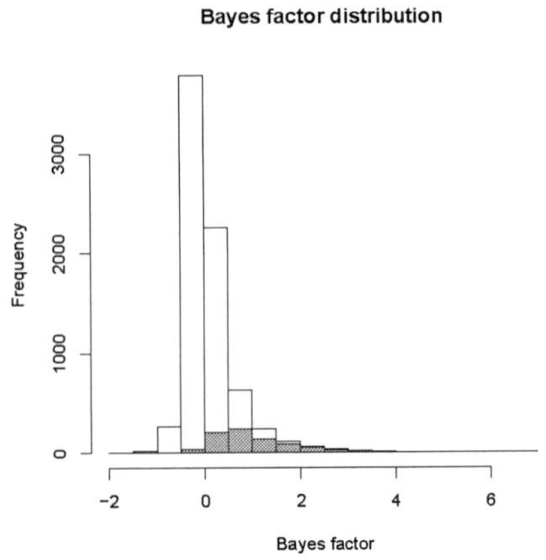

Fig. 3. Distribution of Bayes factors for the 7144 genes on the muscular dystrophy dataset. Shaded in grey are the Bayes factors for the genes with p-value less than 0.05 from the Hotelling T^2-test.

based pathway analysis. To avoid a bias due to the different length of the gene lists, we compared the 96 genes selected by the Bayesian test with the top 100 genes selected by the Hotelling T^2-test. With a recently published text mining tool for the analysis of Medline abstract [12], we analyzed whether the selected genes were mentioned in the same abstract as the term "muscular dystrophy" or a related term. From the list selected by the Bayesian test, 28% of the genes is mentioned together with "muscular dystrophy", compared to 18% of the genes out of the top 100 from the Hotelling test. Notably, also 5 out of the 16 genes selected only by the Bayesian method have been previously associated with muscular dystrophy. Since not all the relationships between genes and the process of muscular dystrophy have been made explicit in literature, we also matched the literature fingerprints of these genes (i.e. weighted lists of biological concepts associated with these genes [12]), with the literature fingerprint of "muscular dystrophies". We found a stronger association of the genes selected by the Bayesian method with the concept "muscular dystrophies": the average match score for these genes was 0.023 compared to 0.0058 for the top 100 genes from the Hotelling T^2-test. We have also analyzed which biological pathways (as defined by the Gene Ontology consortium, www.geneontology.org) were overrepresented in the lists of genes identified by the two methods, using the web-based tool Webgestalt (http://bioinfo.vanderbilt.edu/webgestalt). Several pathways relevant to the muscular phenotype were found significant at a p-value less than 0.05 (using a hypergeometric test) for the list of genes coming from the Bayesian test: uscle contraction, immune response, chemokine activity, and sarcomere genes. These pathways nor other relevant pathways were significantly overrepresented in the top 100 of genes from the Hotelling T^2-test. This confirms that the newly developed Bayesian test indeed prioritises genes with relevance to the muscular dystrophy process.

In addition, we have looked at the 6 genes that were selected for biological validation in [1] and on which also qPCR experiments were conducted. On these, we found a very similar conclusion to the one obtained with the Hotelling T^2-test, with *Dlk1*,*Casq2* and *Dpp4* having a high Bayes factor (greater than 2.1) and *Tcap*, *Dbp* and *Myoz2* having a low Bayes factor (less than 0.4). Furthermore, despite the restriction on the quadratic term of the polynomial adopted in this application, the test has managed to pick up with strong evidence 7 out of the 11 genes found by the Hotelling test with an extreme quadratic coefficient for the temporal profile and a further 2 with moderate evidence. These genes are supported by biological knowledge [1]. When considering the 2 genes that were picked up by the Hotelling test but not by the Bayesian test and comparing them with the remaining 9 genes, we noticed a smaller difference between the temporal profiles across the conditions with respect to the associated noise. This is probably the reason why these genes have not been picked up by the Bayesian test and shows that the test is less sensitive to noise.

4 Conclusion

In this paper, we have proposed a novel Bayesian median regression model for temporal gene expression data. A quadratic polynomial was used to fit the temporal profile of a gene under a specific biological condition. A full MCMC Bayesian inference is proposed, with the aim of increasing the robustness of the model to noise and small sample size. A simple Bayes factor test is finally considered to detect the genes whose temporal profile is significantly different across the biological conditions.

We have compared the method, both on simulated and real data, with a classical Hotelling T^2-test recently proposed to tackle the same problem [1]. Promising results have shown the potential of this methodology on the analysis of gene expression data, whereby a relatively small number of genes was detected with strong evidence for differentially expression. Some of these genes had already been biologically verified and a significant number was found with relevance to the muscular dystrophy process using a literature-based approach and a GeneOntology-basd pathway analysis.

The work in this paper can be extended in different directions. Firstly, taking advantage of the increased robustness of the method, more complex temporal models can be considered than a quadratic polynomial, such as polynomials of higher degree or autoregressive models. Secondly, the Bayesian framework of the method means that different prior distributions can be considered that might be better suited to this application. Thirdly, the same procedure described in this paper can be used to estimate quantiles different than the median, as considered for example by [2]. In this way, a more complete picture of the gene expression distribution can be obtained and decisions can be made based on this. Finally, the approach does not make any assumption on the error distribution and as such it can be applied on a variety of applications.

Acknowledgement

The work of V.V. on this paper was funded by the BBSRC (Grant: EGM17735).

References

1. Vinciotti, V., Liu, X., Turk, R., de Meijer, E., 't Hoen, P.A.: Exploiting the full power of temporal gene expression profiling through a new statistical test: Application to the analysis of muscular dystrophy. BMC Bioinformatics **7**(183) (2006)
2. Wang, H., He, X.: Detecting differential expressions in GeneChip microarray studies: a quantile approach. JASA **102**(477) (2007) 104–112
3. Lewin, A., Richardson, S., Marshall, C., Glazier, A., Aitman, T.: Bayesian modeling of differential gene expression. Biometrics **62**(1) (2006) 10–18
4. Yu, K., Moyeed, R.: Bayesian quantile regression. Statistics and Probability Letters **54** (2001) 437–447
5. Yu, K., Stander, J.: Bayesian analysis of a tobit quantile regression model. Journal of Econometrics **137** (2007) 260–276
6. Koenker, R., Bassett, G.: Regression quantiles. Econometrica **46**(1) (1978) 33–50
7. Bernardo, J., Smith, A.: Bayesian Theory. John Wiley & Sons, Chichester (2004)
8. Kass, R.: Bayes factors. JASA **90** (1995) 773–795
9. Jeffreys, H.: The theory of probability. The Oxford University Press (1961)
10. Wasserman, L.: Bayesian model selection and model averaging. Journal of Mathematical Phycology **44** (2000) 92–107
11. Vinciotti, V., Khanin, R., D'Alimonte, D., Liu, X., Cattini, N., Hotchkiss, G., Bucca, G., de Jesus, O., Rasaiyaah, J., Smith, C., Kellam, P., Wit, E.: An experimental evaluation of a loop versus a reference design for two-channel microarrays. Bioinformatics **21**(4) (2005) 492–501
12. Jelier, R., Jenster, G., Dorssers, L., Wouters, B., Hendriksen, P., Mons, B., R., D., Kors, J.: Text-derived concept profiles support assessment of DNA microarray data for acute myeloid leukemia and for androgen receptor stimulation. BMC Bioinformatics **8**(14) (2007)

SYSTEMS BIOLOGY

Minimal Direction Cuts in Metabolic Networks

Abdelhalim Larhlimi and Alexander Bockmayr

DFG-Research Center Matheon, FB Mathematik und Informatik, Freie Universität Berlin, Arnimallee, 3, 14195 Berlin, Germany

e-mail: larhlimi@mi.fu-berlin.de,bockmayr@mi.fu-berlin.de

Abstract. Minimal cut sets (MCSs) have been introduced to define perturbations in a metabolic network needed to repress the activity of a given target reaction. While this approach gives interesting insight into the structure of metabolic networks, it suffers from two major drawbacks that are limiting in practice. First, using MCSs becomes inadequate when we are interested in inhibiting only one direction of a reversible target reaction. Second, and more importantly, the computation of MCSs by currently available methods is limited by the combinatorial explosion of the number of elementary modes and hence is not practical for genome-scale networks. To overcome these problems, we introduce the concept of minimal direction cuts. A minimal forward (resp. backward) direction cut (MFC) (resp. MBC) is a minimal set of reactions whose removal prevents the target reaction from carrying a flux in the forward (resp. backward) direction. Computing MFCs and MBCs does not require to determine beforehand the elementary modes of the network. Instead it can be based on the Farkas lemma for equality and inequality constraints. This leads to a new algorithm to identify minimal direction cuts, which can also be used for a direct computation of MCSs.

1 Introduction

Constraint-based approaches have become a central tool to analyze genome-scale metabolic networks [9]. These approaches analyse the possible behaviors of a metabolic network given stoichiometric and thermodynamic constraints. Taken together, these constraints not only limit the allowable flux distributions over the network at steady state. They also indicate the level of importance of individual reactions in the metabolism. Several approaches have been proposed to assess the importance of reactions [2, 3, 8]. These approaches not only help us to enhance our understanding of biological systems. They also can be used to determine a reaction that is critical for the survival of a certain pathogen. Such a reaction could be a suitable target to be repressed by network perturbations. A recent approach [4, 5] has introduced minimal cut sets to identify such network perturbations. A *minimal cut set* (MCS) is a minimal set of reactions whose removal represses a given target reaction. However, this approach suffers from two major drawbacks that are limiting in practice. First, using MCSs becomes inadequate when we are interested in inhibiting only one direction of a reversible target reaction. Second, and more importantly, the computation of MCSs is

CP940, *CompLife 2007, 3rd International Symposium on Computational Life Science*
edited by A. Siebes, M. Berthold, R. Glen, and A. Feelders

currently based on the conservation property of elementary modes (EMs) [3, 11] and the principle that each MCS is a minimal hitting set for all EMs involving the target reaction. This computation suffers from the combinatorial explosion of the number of EMs [6] and hence becomes impractical for genome-scale networks.

To overcome these limitations, we introduce the concept of *minimal direction cuts* in metabolic networks. A *minimal forward* (resp. *backward*) *direction cut (MFC)* (resp. *MBC*) is a minimal set of reactions whose removal prevents the target reaction from carrying a flux in the forward (resp. backward) direction. If the target reaction is irreversible, MFCs are identical to MCSs and the empty set is the unique MBC. If, however, the target reaction is reversible, each MCS is the union of an MFC and an MBC. More importantly, the computation of MFCs and MBCs does not require that of EMs. Instead it can be based on the Farkas lemma for equality and inequality constraints. Finally, MCSs can be directly calculated from MFCs and MBCs.

The organization of this paper is as follows. In Sect. 2, we give a first presentation of our approach. In Sect. 3, we formally characterize MFCs and MBCs using an extended Farkas lemma. In Sect. 4, we propose a recursive algorithm to identify MFCs and MBCs in a metabolic network. Finally, some computational results are given in Sect. 5.

2 Minimal forward and backward direction cuts

In the context of metabolic pathway analysis, metabolic systems are assumed to operate at steady state such that for all internal metabolites the flux is balanced. In addition, the flux through each irreversible reaction must be non-negative. The set of all possible flux distributions through the network at steady state, defines the *flux cone* [9]

$$F = \{v \in \mathbb{R}^n \mid Sv = 0,\ v_i \geq 0,\ \text{for all } i \in \mathit{Irr}\} \tag{1}$$

where S is the $m \times n$ stoichiometric matrix of the network, with m internal metabolites (rows) and n reactions (columns), and $v \in \mathbb{R}^n$ is the *flux vector*. $\mathit{Irr} \subset \{1, \ldots, n\}$ denotes the set of *irreversible* reactions in the network, and $\mathit{Rev} = \{1, \ldots, n\} \setminus \mathit{Irr}$ the set of *reversible* reactions.

For the rest of the paper, let $\tau \in \{1, \ldots, n\}$ be the target reaction. A *forward* (resp. *backward*) *direction cut* for reaction τ is a set of reactions whose removal from the network implies that all possible steady state flux distributions over the network do not involve reaction τ in the forward (resp. backward) direction. Here, the removal of a set of reactions is mathematically expressed by constraining the fluxes through all these reactions to zero.

Definition 1. *Let F be the flux cone defined in equation (1).*

- *A forward direction cut (FC) for reaction τ is a set of reactions $M \subseteq \{1, \ldots, n\}$, such that for any $v \in F$ with $v_i = 0$ for all $i \in M$, $v_\tau \leq 0$.*
- *A backward direction cut (BC) for reaction τ is a set of reactions $M \subseteq \{1, \ldots, n\}$, such that for any $v \in F$ with $v_i = 0$ for all $i \in M$, $v_\tau \geq 0$.*

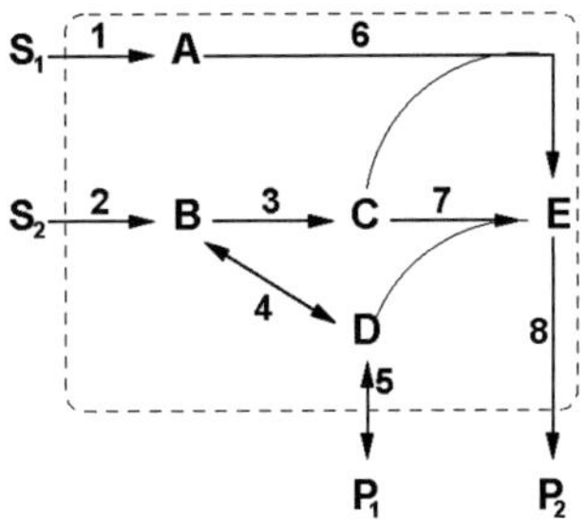

Fig. 1. Network example (ILLUSNET)

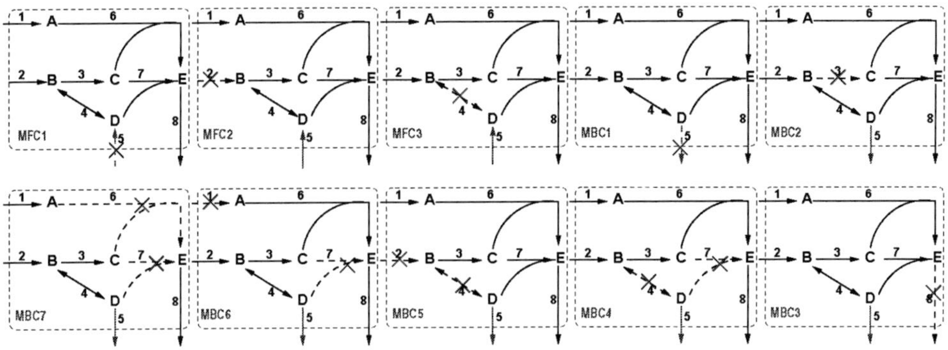

Fig. 2. Minimal forward and backward direction cuts for the target reaction 5.

A forward (resp. backward) direction cut M is minimal, denoted MFC (resp. MBC), if there is no forward (resp. backward) direction cut $M' \subsetneq M$ strictly contained in M.

Example 1. Consider the hypothetical network ILLUSNET [4] depicted in Fig. 1. It consists of five internal metabolites $(A, \ldots, E)$, and eight reactions $(1, \ldots, 8)$, whereof reactions 4 and 5 are reversible. Let reaction 5 be the target reaction. Fig. 2 (resp. Fig. 3) displays MFCs and MBCs (resp. MCSs) for reaction 5. For instance, the forward direction cut MFC2 does not inhibit the backward direction of reaction 5. MCS2 $= \{2, 3\}$ is a minimal cut set to repress reaction 5. However, this set is not minimal to repress the production of metabolite P_1 since this is possible by removing only reaction 2. Indeed, MCS2 is the union of the minimal direction cuts MFC2 and MBC2.

The meaningful direction cuts are those which are minimal, since the trivial solution, which consists of removing all reactions, always is a direction cut for τ. Obviously, if there is no flux vector $v \in F$ with $v_\tau < 0$ (resp. $v_\tau > 0$), then the empty set is the unique MBC (resp. MFC) for τ. In particular, if τ is an irreversible reaction, then the empty set is the unique MBC for τ and MFCs for τ are identical to minimal cut sets. On the other hand, if τ is reversible, then the computation of MFCs is equivalent to that of MBCs in the sense that an

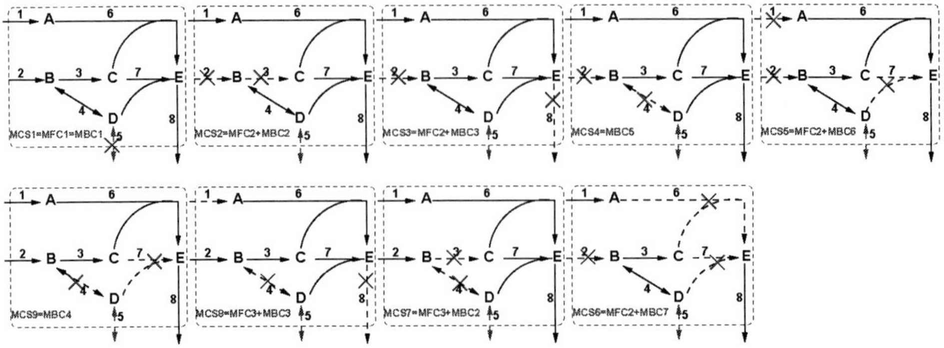

Fig. 3. Minimal cut sets for the target reaction 5.

algorithm for computing MFCs could also be used to calculate MBCs. To show this, let S' be the $m \times n$ matrix whose columns are defined by

$$S'_{*j} = S_{*j} \text{ for all } j \in \{1, \dots, n\} \setminus \{\tau\} \text{ and } S'_{*\tau} = -S_{*\tau},$$

and let

$$F' = \{v \in \mathbb{R}^n \mid S'v = 0,\ v_i \geq 0, \text{ for all } i \in Irr\}.$$

Now one can easily prove the following lemma.

Lemma 1. *Let $M \subseteq \{1, \dots, n\} \setminus \{\tau\}$ be a set of reactions. If τ is reversible, i.e., $\tau \in Rev$, then the following are equivalent:*

1. *For any $v \in F$ with $v_i = 0$ for all $i \in M$, $v_\tau \geq 0$.*
2. *For any $v \in F'$ with $v_i = 0$ for all $i \in M$, $v_\tau \leq 0$.*

Given a set $M \subseteq \{1, \dots, n\} \setminus \{\tau\}$ of reactions, it follows from Lemma 1 that M is an MBC if and only if M is an MFC with respect to the stoichiometric matrix S'. This is why we focus in the following on characterizing MFCs. The same results hold also for MBCs. Note that $\{\tau\}$ is a trivial MFC for τ. Therefore, each MFC $M = \{\tau\}$, called *non-trivial* MFC for τ, does not include τ, i.e., $\tau \notin M$. In the following, we are interested in finding these non-trivial MFCs.

In analogy with minimal cut sets, MFCs could be computed using elementary modes (EMs). An *elementary mode* corresponds to a steady state flux distribution that involves a minimum set of reactions [7, 11]. Whenever the flux through some reaction is constrained to zero, EMs that remain are those which do not contain that reaction. Based on this conservation property, each MFC is then a minimal hitting set of all EMs involving the target reaction in the forward direction. A set $M \subseteq \{1, \dots, n\}$ is a *hitting set* of a collection C of EMs if M meets every EM in the collection C, i.e., $M \cap \{i \mid e_i = 0\} = \emptyset$ for each $e \in C$. Since the computation of EMs often leads to a combinatorial explosion [9], computing MFCs using EMs does not scale well for genome-scale metabolic networks. In the following, we instead use the Farkas lemma for equality and inequality constraints to further characterize MFCs.

3 Characterizing MFCs using the Farkas lemma

In this section we give a formal characterization of minimal forward direction cuts in a metabolic network. We use mainly the Farkas lemma, which states that the unsolvability of a system of constraints can be established by finding a solution for a corresponding dual system.

Theorem 1 (Farkas lemma for equality and inequality constraints [13]). *For compatible matrices A, B, C and vectors x, y, z* either *there exists a solution vector v for*

$$Av = x,\ Bv \geq y,\ Cv \leq z$$

or *there exist row vectors a, b, c with*

$$cC = aA + bB,\ b \geq 0,\ c \geq 0,\ -ax - by + cz < 0$$

Let $M \subseteq \{1, \ldots, n\}$ be an FC and let $(e_1, \ldots, e_n)$ be the canonical basis of $\mathbb{R}^n$. In the following, I_M (resp. I_{Irr}) denotes the matrix whose rows are the vectors e_j with $j \in M$ (resp. $j \in Irr$). From Definition 1, we get that

$$\text{there is no } v \in \mathbb{R}^n \text{ such that } Sv = 0,\ I_M v = 0,\ I_{Irr} v \leq 0 \text{ and } v_\tau \leq -1. \quad (2)$$

Let A, B and C be the matrices defined by

$$A = \begin{pmatrix} S \\ -I_M \end{pmatrix}, B = -I_{Irr} \text{ and } C = e_\tau,$$

and let $x = y = 0$ and $z = -1$. By (2), there exists no solution vector $v \in \mathbb{R}^n$ for $Av = x,\ Bv \geq y,\ Cv \leq z$. According to the Farkas lemma, there exist row vectors $\alpha \in \mathbb{R}^m$, $\mu \in \mathbb{R}^{|M|}$, $\beta \in \mathbb{R}^{|Irr|}$ and $\gamma \in \mathbb{R}$, such that $\gamma C = (\alpha, \mu)A + \beta B$, $\beta \geq 0$, $\gamma \geq 0$ and $\gamma z < 0$. This leads to the following proposition which characterizes FCs using the Farkas lemma.

Proposition 1. *Let $M \subseteq \{1, \ldots, n\}$ be a set of reactions. Then, the following are equivalent:*

1. *M is an FC for τ.*
2. *There exist $\alpha_i, \beta_j, \mu_k, \gamma \in \mathbb{R}$ with $\beta_j \geq 0$ and $\gamma \geq 1$, such that*

$$\sum_{k \in M} \mu_k e_k + \gamma e_\tau = \sum_{i=1}^{m} \alpha_i S_{i*} - \sum_{j \in Irr} \beta_j e_j. \quad (3)$$

Let D be the cone defined by

$$D = \{\sum_{i=1}^{m} \alpha_i S_{i*} - \sum_{j \in Irr} \beta_j e_j \mid \beta_j \geq 0 \text{ for all } j \in Irr\}, \quad (4)$$

and for each vector $w \in D$, let

$$supp(w) = \{i \in \{1, \cdots, n\} \mid w_i \neq 0\} \quad (5)$$

be its support. The following corollary describes FCs by means of *supp* and D.

Corollary 1. *Let D be the cone defined in (4) and let $M \subseteq \{1,\ldots,n\} \setminus \{\tau\}$ be a set of reactions. Then, the following are equivalent:*

1. *M is an FC for τ.*
2. *There exists a vector $w \in D$, such that $w_\tau \geq 1$ and $supp(w) \setminus \{\tau\} \subseteq M$.*

Note that the minimality of MFCs is quite similar to that of elementary modes (EMs). An EM corresponds to a flux vector $e \in F$ involving a minimum set of reactions [11], i.e., $supp(e)$ is minimal. In the same way, an MFC corresponds to a vector $w \in D$ such that $w_\tau \geq 1$ and $supp(w)$ is minimal. We will call such a vector *simple in D with respect to τ and supp*. More generally, we define:

Definition 2. *For $Q \subseteq \mathbb{R}^n$ let $\varphi : Q \to 2^{\{1,\ldots,n\}}$ be a function that maps each vector $s \in Q$ to a subset $\varphi(s) \subseteq \{1,\ldots,n\}$. A vector $s \in Q$ is simple in Q with respect to τ and φ if the following conditions hold:*

1. *$s_\tau \geq 1$.*
2. *There is no vector $s' \in Q$ such that $s'_\tau \geq 1$ and $\varphi(s') \subsetneq \varphi(s)$.*

If this is the case, we also say that $simp_{(Q,\tau,\varphi)}(s)$ holds.

The following theorem shows that MFCs are in a 1-1 correspondence with the support of vectors simple in D with respect to τ and $supp$.

Theorem 2. *Let D be the cone defined in equation (4) and let $M \subseteq \{1,\ldots,n\} \setminus \{\tau\}$ be a set of reactions. Then, the following are equivalent:*

1. *M is an MFC for τ.*
2. *There exists a vector w simple in D with respect to τ and supp, such that $supp(w) \setminus \{\tau\} = M$.*

Next we derive an alternative characterization. Let

$$E = \{\sum_{i=1}^{m} \alpha_i S_{i*} \mid \alpha_i \in \mathbb{R} \text{ for all } i = 1,\ldots,m\}, \tag{6}$$

and let $supp' : E \to 2^{\{1,\ldots,n\}}$ be the function that maps each vector $u \in E$ to

$$supp'(u) = \{i \in Irr \setminus \{\tau\} \mid u_i < 0\} \cup \{i \in Rev \setminus \{\tau\} \mid u_i \neq 0\} \tag{7}$$

Suppose $w = \sum_{i=1}^{m} \alpha_i S_{i*} - \sum_{j \in Irr} \beta_j e_j$ is simple in D with respect to τ and $supp$, and let $s = \sum_{i=1}^{m} \alpha_i S_{i*}$. One can easily show that $supp(w) \setminus \{\tau\} = supp'(s)$. More importantly, s is simple in E with respect to τ and $supp'$ (see Appendix A). Therefore, MFCs can be identified using vectors simple in E with respect to τ and $supp'$. This reduces the complexity of the MFC computation since E is generated with only m generators, while D is spanned by $m + |Irr|$ generators and defined by $|Irr|$ additional constraints.

Theorem 3. *Let E be the cone defined in (6) and let $M \subseteq \{1,\ldots,n\} \setminus \{\tau\}$ be a set of reactions. Then, the following are equivalent:*

1. *M is an MFC for* τ.
2. *There exists a vector* s *simple in* E *with respect to* τ *and* $supp'$ *such that* $supp'(s) = M$.

The results above show that MFCs can be identified using vectors simple in E with respect of τ and $supp'$. These results can also be used to determine MCSs. Indeed, an MCS can be seen as the union of an MFC and an MBC depending on the reversibility type of the target reaction. If the latter is irreversible, the empty set is the unique MBC and MFCs are identical to MCSs.

Corollary 2. *Let* $M \subseteq \{1, \ldots, n\} \setminus \{\tau\}$ *be a set of reactions. If reaction* τ *is irreversible, i.e.,* $\tau \in Irr$*, then the following are equivalent:*

1. *M is an MCS for* τ.
2. *M is an MFC for* τ.

The following corollary shows that, if the target reaction is reversible, each MCS is the union of an MFC and an MBC.

Corollary 3. *Let* $M \subseteq \{1, \ldots, n\} \setminus \{\tau\}$ *be a set of reactions. If reaction* τ *is reversible, i.e.,* $\tau \in Rev$*, and* M *is an MCS for* τ*, then there exist an MFC* M_1 *and an MBC* M_2 *for* τ*, such that* $M = M_1 \cup M_2$.

The mathematical results above suggest an algorithm to identify MFCs by means of vectors simple in E with respect of τ and $supp'$. One possible approach to determine these simple vectors is the use of mixed-integer linear programming (MILP).

4 Algorithm for finding MFCs

We have shown that each MFC corresponds to a simple vector in E with respect to τ and $supp'$. To identify such a simple vector, we first introduce for each reaction $j = \tau$ a binary variable λ_j. By L and U we denote resp. some lower and upper bound for vectors $s \in E$, i.e., $L \leq s_j \leq U$ for all $j \in \{1, \ldots, n\} \setminus \{\tau\}$. A simple vector in E with respect to τ and $supp'$ corresponds to an optimal solution of the following MILP:

$$
\begin{array}{lll}
\min & \sum_{j=\tau} \lambda_j & \\
\text{s. t.} & s_\tau \geq 1 & \\
 & L\lambda_j \leq s_j \leq U\lambda_j, \quad \forall j \in Rev, & \text{(a)} \\
 & L\lambda_j \leq s_j, \quad \forall j \in Irr \setminus \{\tau\}, & \text{(b)} \\
 & (s_1, \ldots, s_n) \in E, & \\
 & \lambda_j \in \{0, 1\}, \quad \forall j \in \{1, \ldots, n\} \setminus \{\tau\} &
\end{array}
$$

Conditions (a) and (b) ensure for all $j \in \{1, \ldots, n\} \setminus \{\tau\}$ that $j \in supp'(s)$ implies $\lambda_j = 1$. Although a solution (s, λ) may exist with $\lambda_k = 1$ and $k \notin supp'(s)$ for some $k \in \{1, \ldots, n\} \setminus \{\tau\}$, this solution is not optimal. Indeed, (s, λ') with $\lambda'_k = 0$

and $\lambda'_j = \lambda_j$ for all $j \neq k$ would be another solution with a smaller value of the objective function. For an optimal solution (s^*, λ^*), we have

$$\lambda_j^* = 1 \text{ if and only if } j \in \mathit{supp}'(s^*). \tag{8}$$

Therefore, s^* is a simple vector in E with respect to τ and supp', and $M = \mathit{supp}'(s^*) = \{j \mid \lambda_j^* = 1\}$ is an MFC.

Here, we are interested in finding all MFCs. We propose a recursive MILP that obtains, in each iteration i, an optimal solution (s^*, λ^*) and so an MFC $M^i = \{j \mid \lambda_j^* = 1\}$. We ensure the selection of a new MFC M^i by imposing the following constraint on the binary variables:

$$\sum_{j \in M^{i-1}} \lambda_j \leq |M^{i-1}| - 1 \tag{9}$$

where M^{i-1} is the MFC obtained in the previous iteration $i-1$. This condition signifies that at least one reaction $j \in M^{i-1}$ is not included in M^i, otherwise M^{i-1} would be contained in M^i, and since the latter is minimal, we would get $M^i = M^{i-1}$. The algorithm stops when no other optimal solution can be found.

Flux coupling analysis [1, 8] could be used to optimize our algorithm. This analysis allows identifying directional couplings between reactions. A reaction j is directionally coupled with some reaction i, written $i \overset{=0}{\rightarrow} j$, if a zero flux through i implies a zero flux through j. In this case, at most i or j is contained in an MFC. This is expressed by the following constraint:

$$\lambda_i + \lambda_j \leq 1 \quad \forall i, j \in \{1, \ldots, n\} \setminus \{\tau\} \text{ such that } i \overset{=0}{\rightarrow} j \tag{10}$$

In addition, a reaction i such that $i \overset{=0}{\rightarrow} \tau$ is itself a trivial MFC. Therefore, the following constraint holds for non-trivial MFCs:

$$\lambda_i = 0 \quad \forall i \in \{1, \ldots, n\} \setminus \{\tau\} \text{ such that } i \overset{=0}{\rightarrow} \tau \tag{11}$$

A typical example of coupled reactions are *enzyme subsets* [10]. Indeed, the fluxes through reactions in an enzyme subset are proportional to each other. Given the enzyme subsets of a metabolic network, we can obtain, in a preprocessing step of our algorithm, a compressed network with the enzyme subsets taken as combined reactions. This compression, which is is identical to that presented in [3], results in a smaller network, and so computing the MFCs for the compressed network is easier than computing those of the original network.

Our approach may be hampered by the huge number of possible minimal direction cuts. However, with our method, we can easily add new constraints to reduce the search space. For instance, from the practical viewpoint, we could require that minimal direction cuts contain only external reactions. These reactions reflect the experimental conditions the biological system is placed on and so could be easy to control experimentally. On the other hand, while looking for minimal direction cuts for a target reaction, we should be careful not to perturb involuntarily other reactions that are important for the whole metabolism.

Input : Target reaction τ;
Stoichiometric matrix S;
Set of irreversible reactions $Irr \subseteq \{1, \ldots, n\}$;
Maximal size of MFCs MaxSize;

Optional: Coupled reactions
$\Phi = \{(i,j) \mid i \stackrel{=0}{\rightarrow} j \vee j \stackrel{=0}{\rightarrow} i, 1 \leq i < j \leq n, i \neq \tau, j \neq \tau\}$;
$\Psi = \{i \mid i \stackrel{=0}{\rightarrow} \tau, 1 \leq i \leq n, i \neq \tau\}$;
Set $\Delta \subseteq \{1, \ldots, n\}$ of reactions that must not occur in MFCs.

Output : Set F of minimal forward direction cuts (MFCs).

Initialization: $M := \emptyset$,

```
/* Trivial MFCs */
```

add($\{\tau\}, F$),
foreach $i \in \Psi$ **do**
  if $i \notin \Delta$ **then**
    add($\{i\}, F$));
  end
end
repeat

```
/* Find a vector simple in E with respect to τ and supp' */
```

$$
\begin{aligned}
(\alpha^*, \lambda^*) = \min\ & \textstyle\sum_{j \neq \tau} \lambda_j \\
\text{s. t. } & \textstyle\sum_{i=1}^{m} \alpha_i S_{i\tau} \geq 1 \\
& L\lambda_j \leq \textstyle\sum_{i=1}^{m} \alpha_i S_{ij} \leq U\lambda_j \quad \forall j \in Rev, \\
& L\lambda_j \leq \textstyle\sum_{i=1}^{m} \alpha_i S_{ij} \quad \forall j \in Irr \setminus \{\tau\}, \\
& \lambda_i + \lambda_j \leq 1 \quad \forall (i,j) \in \Phi, \\
& \lambda_i = 0 \quad \forall i \in \Psi, \\
& \lambda_i = 0 \quad \forall i \in \Delta, \\
& \textstyle\sum_{j \neq \tau} \lambda_j \leq \text{MaxSize} \\
& \lambda_j \in \{0, 1\} \quad \forall j \in \{1, \ldots, n\} \setminus \{\tau\}, \\
& \alpha_i \in \mathbb{R} \quad \forall i \in \{1, \ldots, m\}.
\end{aligned}
$$

```
/* Retrieve the corresponding MFC */
```

$M = \{j \mid \lambda_j^* = 1\}$;
add(M, F),

```
/* Force the selection of a new MFC in the next iteration */
```

add to the constraint system: $\sum_{j \in M} \lambda_j \leq |M| - 1$.

until *No other solution is found* ;

Algorithm 1: Minimal Direction Cut Finder

As an example, consider again the network depicted in Fig. 1. Assume we wish to repress the consumption of metabolite P_1 and maintain the production of metabolite P_2. Removing reaction 3 inhibits the consumption of P_1. However, this perturbation forbids also the production of P_2. Therefore, reaction 3 should be excluded from minimal direction cuts. In general, we could define a set of reactions $\Delta \subseteq \{1, \ldots, n\}$ that should not occur in minimal direction cuts, and then impose the following additional condition on the binary variables corresponding to these reactions:

$$\lambda_i = 0 \quad \forall i \in \Delta. \tag{12}$$

Finally, we often are interested in removing a small number of reactions from a metabolic network. Enumerating all the MFCs is not always necessary. Our algorithm allows for defining the maximal size (MaxSize) of the MFCs to be computed by imposing the following constraint:

$$\sum_{j \neq \tau} \lambda_j \leq \text{MaxSize} \tag{13}$$

5 Computational results for the central metabolism of E. coli

Our recursive algorithm, whose main steps are shown in Algorithm 1, has been implemented in the Java language. We used CPLEX 9.0 as our solver for linear and integer programming problems, accessed via Java. In order to check the capabilities of our algorithm, we computed the MFCs in the central metabolic network of *E. coli* with 'biomass reaction' as a target reaction. For our computations, we considered four variants of the *E. coli* central metabolism corresponding to growth on four different substrates (acetate, succinate, glycerol and glucose) [12]. In order to work on networks with different complexities, several pseudo reactions have been inserted in these variants. The number of elementary modes is somehow a measure indicating the degree of complexity of analyzing the network. This observation is important because the difficulty of computing MCSs using the traditional algorithm grows with the number of elementary modes. On the other hand, the target reaction is irreversible, and so the MFCs and MCSs are identical. Therefore, working on these four variants also allows for comparing a direct computation of MCSs using our algorithm with a computation using elementary modes. Experiments for computing MCSs in the four variants of the *E. coli* metabolism using the traditional algorithm are described in [5].

Table 1 summarizes our computational results. The computations were performed on a Solaris server with a Sparc III Processor 900 MHz and 32 GB RAM. We did not use additional constraints (10), (11) and (12) that would allow for reducing the search space. The computation times of our current prototype implementation may be compared with those given in [5]. Note that the latter do not include the time needed for computing the elementary modes, which is known to be a complex combinatorial problem [6]. In addition, our results show that the difficulty of computing MFCs grows with the number of elementary modes. This observation, which holds also for MCSs, reflects that the number of elementary modes is a measure for the complexity of the network. Note however that, contrary to MCSs, the computation of MFCs does not require that of elementary modes.

Table 1. Computation of MFCs for variants of the central metabolism of *E. coli*, with the number of reactions (n), the number of metabolites (m), the number of elementary modes (♯EMs) and the number of MFCs (♯MFCs). Contrary to MCSs, the computation of MFCs does not require that of elementary modes. For all computations, we defined six as the maximal size of the MFCs.

Substrate	Acetate	Succinate	Glycerol	Glucose
♯reactions (n)	104	104	106	105
♯metabolites (m)	88	88	89	88
♯EMs (network complexity)	599	4250	11333	27100
♯MFCs	206	340	624	1249
Computation time	1min 28s	22m 20s	1h 14min	12h 10min

6 Conclusion

In this paper, we have introduced minimal direction cuts (MDCs) that allow preventing a target reaction from being performed in an undesired direction. Obviously, if the target reaction is irreversible, MDCs are not different from minimal cut sets (MCSs). However, if this reaction is reversible, MDCs allow it to operate in the desired direction, while MCSs make it inactive in both directions. In both cases, each MCS can be seen as the union of two MDCs. Therefore, all the useful applications of MCSs [5] can also done with MDCs. However, the computation of MDCs does not require that of elementary modes. MDCs may be determined based on the Farkas lemma for equality and inequality constraints. The mathematical results of this paper have led to a recursive algorithm for computing MDCs in a metabolic network. Since MCSs can be obtained from MDCs, this algorithm also allows for a direct computation of MCSs, in the sense that we need not calculate beforehand the elementary modes. Finally, our algorithm gives the possibility of introducing additional constraints that may reduce the search space. This makes our approach applicable even for genome-scale metabolic networks.

References

1. A.P. Burgard, E.V. Nikolaev, C.H. Schilling, and C.D. Maranas. Flux coupling analysis of genome-scale metabolic network reconstructions. *Genome Res.*, 14(2):301–312, 2004.
2. A.P. Burgard, S. Vaidyaraman, and C.D. Maranas. Minimal reaction sets for escherichia coli metabolism under different growth requirements and uptake environments. *Biotechnol. Prog.*, 17(5):791–797, 2001.
3. J. Gagneur and S. Klamt. Computation of elementary modes: a unifying framework and the new binary approach. *BMC Bioinformatics*, 5:175, 2004.
4. S. Klamt. Generalized concept of minimal cut sets in biochemical networks. *Biosystems*, 83(2-3):233–247, 2006.
5. S. Klamt and E.D. Gilles. Minimal cut sets in biochemical reaction networks. *Bioinformatics*, 20(2):226–234, 2003.

6. S. Klamt and J. Stelling. Combinatorial complexity of pathway analysis in metabolic networks. *Mol. Bio. Rep.*, 29(1-2):233–236, 2002.
7. S. Klamt and J. Stelling. Two approaches for metabolic pathway analysis? *Trends Biotechnol.*, 21(2):64–69, 2003.
8. A. Larhlimi and A. Bockmayr. A new approach to flux coupling analysis of metabolic networks. In *Computational Life Sciences II, CompLife'06, Cambridge, UK*, pages 205–215. Springer, LNBI 4216, 2006.
9. J.A. Papin, J. Stelling, N. D. Price, S. Klamt, S. Schuster, and B. O. Palsson. Comparison of network-based pathway analysis methods. *Trends Biotechnol.*, 22(8):400–405, 2004.
10. T. Pfeiffer, I. Sánchez-Valdenebro, J.C. Nuno, F. Montero, and S. Schuster. Metatool: for studying metabolic networks. *Bioinformatics*, 15(3):251–257, 1999.
11. S. Schuster, C. Hilgetag, J.H. Woods, and D.A. Fell. Reaction routes in biochemical reaction systems: algebraic properties, validated calculation procedure and example from nucleotide metabolism. *J. Math. Biol.*, 45(2):153–181, 2002.
12. J. Stelling, S. Klamt, K. Bettenbrock, S. Schuster, and E.D. Gilles. Metabolic network structure determines key aspects of functionality and regulation. *Nature*, 420(6912):190–193, 2002.
13. Günter M. Ziegler. *Lectures on Polytopes.* Springer-Verlag, 1994.

Appendix A

Proof of Corollary 1: (1) $\Rightarrow$ (2): By Proposition 1, there exist α_i, β_j, μ_k, $\gamma \in \mathbb{R}$ with $\beta_j \geq 0$, and $\gamma \geq 1$ fulfilling equation (3). Let $w = \sum_{k \in M} \mu_k e_k + \gamma e_\tau$. On the one hand, since $\tau \notin M$, we get $w_\tau = \gamma \geq 1$ and $supp(w) \setminus \{\tau\} = \{k \in M | \mu_k \neq 0\} \subseteq M$. On the other hand, we have $w = \sum_{i=1}^{m} \alpha_i S_{i*} - \sum_{j \in Irr} \beta_j e_j$, $w \in D$ and so the claim follows.

(1) $\Leftarrow$ (2): Since $w \in D$, there exist $\alpha_i, \beta_j \in \mathbb{R}$ with $\beta_j \geq 0$, such that $w = \sum_{i=1}^{m} \alpha_i S_{i*} - \sum_{j \in Irr} \beta_j e_j$. Moreover, since $\tau \notin M$ and $supp(w) \setminus \{\tau\} \subseteq M$, we can write $w = \sum_{k \in M} w_k e_k + w_\tau e_\tau$ with $w_k = 0$ for each $k \in M \setminus supp(w)$. We then get

$$\sum_{k \in M} w_k e_k + w_\tau e_\tau = \sum_{i=1}^{m} \alpha_i S_{i*} - \sum_{j \in Irr} \beta_j e_j.$$

By Proposition 1, M is an FC for τ.

Proof of Theorem 2: (1) $\Rightarrow$ (2): By Corollary 1, there exists $w \in D$ such that $w_\tau \geq 1$ and $supp(w) \setminus \{\tau\} \subseteq M$. Let $N = supp(w) \setminus \{\tau\}$. By Corollary 1, N is an FC for τ. Since N is contained in M and M is minimal, we get $M = N = supp(w) \setminus \{\tau\}$. Suppose that $simp_{(D,\tau,supp)}(w)$ does not hold. Since $w_\tau \geq 1$, there exists a vector $w' \in D$ such that $w'_\tau \geq 1$ and $supp(w') \subsetneq supp(w)$. Let $N' = supp(w') \setminus \{\tau\}$. By Corollary 1, N' is an FC for τ. Moreover, $N' = supp(w') \setminus \{\tau\} \subsetneq supp(w) \setminus \{\tau\} = M$, which is a contradiction.

(1) $\Leftarrow$ (2): By Corollary 1, M is an FC for τ. Suppose that M is not minimal. Then there exists a set N such that N is an FC for τ and $N \subsetneq M$. By Corollary 1, there exists $w' \in D$ such that $w'_\tau \geq 1$, $supp(w') \setminus \{\tau\} \subseteq N$ and so $supp(w') \subsetneq supp(w)$, which is a contradiction.

Lemma 2. *Let D be the cone defined in equation (4) and let $w = \sum_{i=1}^{m} \alpha_i S_{i*} - \sum_{j \in Irr} \beta_j e_j$ be a simple vector in D with respect to τ and supp. Then, for all $k \in Irr \setminus \{\tau\}$, the following are equivalent:*

1. $w_k \neq 0$,
2. $\sum_{i=1}^{j=m} \alpha_i S_{ik} < 0$.

Proof of Lemma 2: (1) $\Leftarrow$ (2): is immediate.

(1) $\Rightarrow$ (2): Suppose that $w_k \neq 0$, $\sum_{i=1}^{j=m} \alpha_i S_{ik} \geq 0$ and let

$$w' = \sum_{i=1}^{m} \alpha_i S_{i*} - \sum_{j \in Irr \setminus \{k\}} \beta_j e_j - (\sum_{i=1}^{m} \alpha_i S_{ik}) e_k$$

We then have $w' \in D$. Since $w' = w - w_k e_k$, $w_k \neq 0$ and $w'_k = 0$, we get $supp(w') \subsetneq supp(w)$. Moreover, since $k \neq \tau$, we get $w'_\tau = w_\tau \geq 1$. Thus, there exists a vector $w' \in D$ such that $w'_\tau \geq 1$ and $supp(w') \subsetneq supp(w)$, which is a contradiction.

Corollary 4. *Let D and E be the cones defined in equations (4) and (6). Let $w = \sum_{i=1}^{m} \alpha_i S_{i*} - \sum_{j \in Irr} \beta_j e_j$ be a simple vector in D with respect to τ and supp and let $s = \sum_{i=1}^{m} \alpha_i S_{i*}$. Then,*

$$supp(w) \setminus \{\tau\} = supp'(s)$$

Lemma 3. *Let D be the cone defined in equation (4). Let $w = \sum_{i=1}^{m} \alpha_i S_{i*} - \sum_{j \in Irr} \beta_j e_j$ be a vector in D and let $s = \sum_{i=1}^{m} \alpha_i S_{i*}$. Then, the following hold:*

1. *$supp'(s) \subseteq supp(w) \setminus \{\tau\}$.*
2. *If $simp_{(D,\tau,supp)}(w)$ holds, then $simp_{(E,\tau,supp')}(s)$ holds.*

Proof of Lemma 3:

1. Let $k \in supp'(s)$. We have then $k \neq \tau$. If $k \in Irr$, then $s_k < 0$, $w_k = s_k - \beta_k < 0$ and so $k \in supp(w) \setminus \{\tau\}$. Otherwise, we get $w_k = s_k \neq 0$ and so $k \in supp(w) \setminus \{\tau\}$. Thus, $supp'(s) \subseteq supp(w) \setminus \{\tau\}$.
2. Suppose that $simp_{(E,\tau,supp')}(s)$ does not hold. Since $s_\tau \geq w_\tau \geq 1$, there exists a vector $s' \in E$ such that $s'_\tau \geq 1$ and $supp'(s') \subsetneq supp'(s)$. Let

$$w' = s' - \sum_{j \in Irr \setminus \{\tau\}, s'_j > 0} s'_j e_j.$$

We have $w' \in D$, $w'_\tau = s'_\tau \geq 1$ and $supp(w') \setminus \{\tau\} = supp'(s') \subsetneq supp'(s)$. Moreover, since $simp_{(D,\tau,supp)}(w)$ holds, it follows from Corollary 4 that $supp'(s) = supp(w) \setminus \{\tau\}$. Thus, there exists a vector $w' \in D$ with $w'_\tau \geq 1$ and $supp(w') \subsetneq supp(w)$, which is a contradiction.

Proof of Theorem 3: (1) $\Rightarrow$ (2) By Theorem 2, there exists a vector $w = \sum_{i=1}^{m} \alpha_i S_{i*} - \sum_{j \in Irr} \beta_j e_j$ simple in D with respect to τ and *supp*, such that $M = supp(w) \setminus \{\tau\}$. Let $s = \sum_{i=1}^{m} \alpha_i S_{i*}$. By Lemma 3, s is simple in E with respect to τ and $supp'$. Moreover, following Corollary 4, $supp'(s) = supp(w) \setminus \{\tau\} = M$.

(2) $\Rightarrow$ (1) Let

$$w = s - \sum_{j \in Irr \setminus \{\tau\}, s_j > 0} s_j e_j.$$

We have $w \in D$, $w_\tau = s_\tau \geq 1$ and $supp(w) \setminus \{\tau\} = supp'(s) = M$. Suppose that $simp_{(D,\tau,supp)}(w)$ does not hold . Then, since $w_\tau \geq 1$, there exists a vector $w' = \sum_{i=1}^{m} \alpha'_i S_{i*} - \sum_{j \in Irr} \beta'_j e_j$ in D, such that $w'_\tau \geq 1$ and $supp(w') \subsetneq supp(w)$. Since $supp(w) \setminus \{\tau\} = supp'(s)$ and $supp(w') \subsetneq supp(w)$, we get $supp(w') \setminus \{\tau\} \subsetneq supp'(s)$. Let $s' = \sum_{i=1}^{m} \alpha'_i S_{i*}$. We have then $s'_\tau \geq w'_\tau \geq 1$. Moreover, by Lemma 3, $supp'(s') \subseteq supp(w') \setminus \{\tau\}$ and so $supp'(s') \subsetneq supp'(s)$, which is a contradiction. Thus, w is simple in D with respect to τ and *supp* and so, following Theorem 2, M is an MFC for τ.

Modeling Evolution of Regulatory Networks in Artificial Organisms

Yolanda Sánchez-Dehesa[1], Guillaume Beslon[1], and José-María Peña[2]

[1] Laboratoire d'InfoRmatique en Images et Systèmes d'information
INSA-Lyon, LIRIS, F-69621 Villeurbanne, France
[2] DATSI, Universidad Politécnica de Madrid, 28660, Spain

Abstract. Regulatory networks are not randomly connected. They are modular, scale-free networks and some motifs distribution is clearly different from random distribution. However, the evolutionary causes and consequences of this specific connectivity are mainly unknown. In this paper we propose Raevol, an integrative model to study the evolution of regulatory networks. While most existing models consider direct evolution of the regulatory network, Raevol integrates a realistic genotype-phenotype mapping where the genome undergo mutations that indirectly modify the genetic network. Moreover, the organisms are selected at the phenotype level (which is produced by the genome *via* the regulation network). Thus, in Raevol, the network only indirectly evolve and it can only be selected if its activity influences the phenotype. We plan to use this model to better understand the network evolution and to study the influence of networks topology on evolution.
keywords: evolutionary algorithm, regulatory networks, protein transcription, network evolution, dynamical properties

1 Introduction

Genetic regulatory networks are a part of many cellular processes and the study of these networks and their evolution can provide us a tool to understand cellular mechanisms. However, these networks cannot be easily studied by "in vitro" tools, due to their kinetic aspect and the impossibility of studying evolution in living systems. So, many authors have estimated the evolutionary mechanism by using bio-informatic tools and models [19,17].

Most of computational models directly address the question of genetic networks evolution, separately from cellular evolution and cellular kinetic. Yet, the genetic network is neither directly selected nor directly subjected to mutations: It doesn't evolve by itself. Genetic networks can only evolve as part of an individual (which can be selected or not) and a genome (on which mutations occur).

In this paper we propose an integrated model of regulatory networks to take into account these two aspects in order to study the evolution of genetic networks. Our model integrates a genome, who's going to undergo mutations, a regulatory network (which evolution will be studied) and an evaluation process based on phenotype characteristics.

CP940, *CompLife 2007, 3rd International Symposium on Computational Life Science*
edited by A. Siebes, M. Berthold, R. Glen, and A. Feelders

2 Genetic Regulation

2.1 Principles

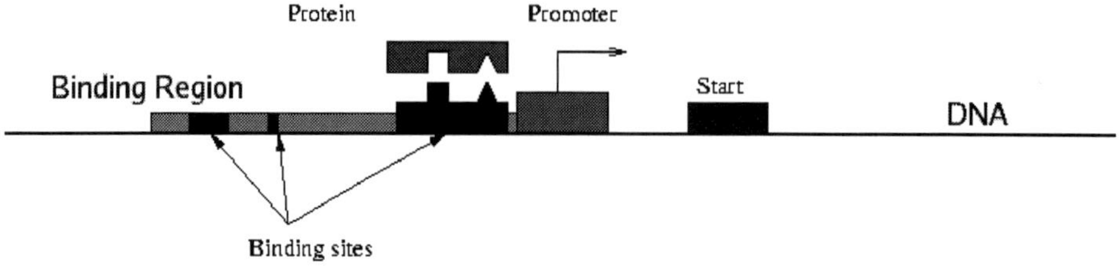

Fig. 1. Example about protein binding over regulatory sites

As it has been formulated by F. Crick in the 1960's, (in the "Central dogma of molecular biology") the DNA decoding is described as a two steps process: transcription (from DNA to RNA) and translation (from RNA to amino-acid sequence). However this process has quickly been extraordinarily complexified by the discovery of the regulatory principles (Jacob, Monod): some proteins can bind on specific DNA locations (Fig 1), interact with them and with the transcription complex, and finally modify the genes transcription close to the binding region.

In the genotype-phenotype mapping, the first step starts with RNA transcription. RNA contains the genetic information transcribed from the genome. By the action of ribosomes, the RNA chain is translated into proteins. The more RNA is produced from a specific gene, the more the corresponding protein will be produced. Thus, the transcription step is of a high importance: it determines the concentration level of the proteins in the cell.

Once produced, proteins can have multiple activities in the cell, for example: enzymatic proteins (for the catalysis in biochemical reactions, such as metabolism), antibodies (active part of the immune system), structural proteins (part of rigid/flexible components of the cell, like actin filaments or membranes) ... Some proteins have a very specific activity: They regulate DNA-transcription itself. Indeed, the transcription process depends on protein-DNA binding properties. Some proteins own a 3-D structure that provides them the capacity of fixation on DNA. When a protein interacts with DNA, in can modify the behavior of the transcription enzymes, thus modifying the transcription level of the genes closed to it. Such proteins are called, "regulatory proteins" or "transcription factors". The DNA sequence in which a transcription factor binds to is named "binding site".

The regulation activity of the transcription factor also depends on the DNA sequence on the binding sites. Most transcription factors bind DNA in a sequence specific manner, but they are usually capable of binding to more than one version of a sequence with more or less probability (affinity). In general, not all the proteins have the same binding site affinity and the transcription level of genes depends on this. When proteins are fixed at the binding sites (in the regulatory region) they will modify the transcription level of the gene which owns the binded region. This means that some proteins could have influence in the transcription (production) of other proteins.

The overall transcription process can be represented as a network (Fig 2). Each node represents a protein and each arc represents the influence of a protein over the transcription of another one. The weight in the arcs define the regulatory level (enhancing or inhibition) between a given protein and the regulated one.

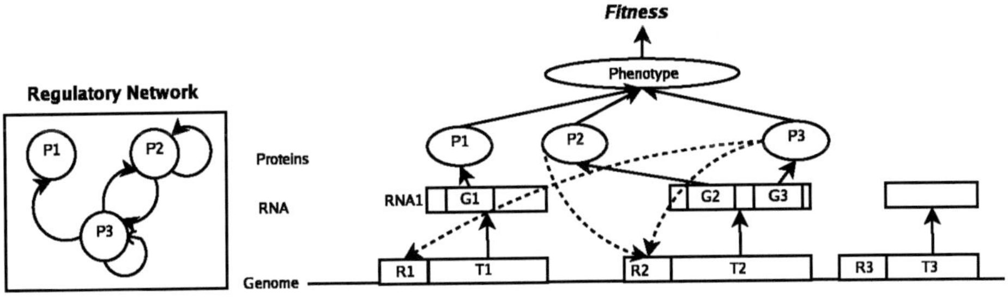

Fig. 2. Regulation in Artificial Evolution. Protein P could be binded in a binding region R, to regulate the transcription of genes G in the T-region

The dynamics of regulatory network is supposed to explain most of the cellular behaviors as cell differentiation, cell response to stress or cell proliferation in cancer. Regulation activity is the core of cell behavior and it could become a key element in many research lines, either in specific cells/organisms or in more general way (e.g. by studying small world properties of the network structure [4]).

The study and modeling of this system is going to provide us a way to analyze development [3] and functional role of proteins over DNA [16], study dynamical properties of cell and to answer several questions about function, expression mechanisms and global interaction [5] in the steady states of cells and also during mitosis.

Regulatory networks, like many other biological networks, are not randomly connected: they follow a power-law distribution [4]. These networks show clustering and modularity phenomena [11] but don't show random distribution. In these networks specific motifs have been detected [10,20]. This connectivity is probably fundamental for the cell activity. Thus, the understanding of both causes and consequences of these specific wiring schemata, is one of the main questions of contemporary biology.

2.2 Biological Regulatory Networks: The lac-operon Example.

A real example of a simple regulatory network can be found in E. Coli bacteria. In this organism's DNA there exists a sequence of consecutive genes, called "lac operon". This sequence indicates how lactose protein compound is metabolized. On the other hand, the lac operon is regulated by several factors, in particular the concentration levels of two proteins, glucose and lactose.

Jacob and Monod experimented with the effects of combinations of sugars as nutrient sources for E. Coli. In the experiments, if glucose and lactose were both provided, glucose would be metabolized first and then lactose. These results come from the regulatory response to high lactose concentration. When this concentration is high RNA polymerase can begin transcription of the operon. Ribosomes then bind to the RNA, and an intermediate enzyme is produced. This enzyme (β-galactosidase) breaks down lactose into two simple sugars (galactose and glucose proteins). When lactose is present but a preferred carbon source (like glucose) is also present then a small amount of enzyme is produced

The general process can be summarized as:

- If lactose concentration is high then lactose is metabolized (lactose concentration is reduced).
- If glucose concentration is high then lactose is not metabolized (lactose concentration remains unchanged).

Another way to explain lac operon function is that lactose has an inhibition influence in lactose concentration, while glucose has an enhancing function on relative lactose concentration.

The speculation of a possible biological explanation is that the metabolism of lactose requires the breakdown of this protein into galactose and glucose, which requires resources (energy to be performed).

E. Coli avoids unnecessary energy consumption metabolizing lactose when an alternative protein (glucose) is available. Glucose is a preferred carbon source because it can be processed directly.

2.3 Evolution of Genetic Networks

As Dobzhansky said 30 years ago: "Nothing in biology makes sense except in the light of evolution". Following this idea, we wonder how networks have evolved to their current topology i.e. what is the evolutionary origin of network structure. It is known that the wiring process has a strong influence on the final network structure and dynamic (See for example preferential attachment from [4]), but the formation process of genetic networks remains mainly unknown.

There exist multiple open issues in the literature of this domain: what's the origin of regulatory networks? How these networks evolve over time? Do they evolve by global duplications or by local tinkering (i.e. one node or one link at a time) ? Studying the inclusion of new nodes in already existing regulatory networks, and studying the development of new regulatory networks could help us to answer some of these questions thus providing us a better understanding on their evolution.

In some specific conditions the modularity phenomenon can appear in evolved regulatory networks [11,9]. However the necessary/sufficient conditions for modularity to appear during evolution are not clear. One can also ask about the role such patterns play during evolution (e.g., do modularity or motifs favor evolution by enabling a better exploration of the functional space ?).

However, for studying network emergence and evolution, we cannot tests our hypothesis on living systems. The solution is to develop a biological model "in silico" that will provide us tools to infer knowledge from artificial network evolution, and to explain how networks evolve.

2.4 Models of Regulatory Networks

There is an impressive amount of literature on the question of regulatory networks modeling. However most of this activity relates to networks kinetic or the inference of genetic regulatory networks from transcriptomic analysis of organisms. As long as evolution of regulatory networks is concerned, authors have mainly focus their works on the question of topology evolution [10,20,11], robustness [6,1,18] and evolution of artificial functions [16,3,2,8]. However most of these models are not realistic enough because they consider direct evolution of the genetic network (i.e. direct encoding of the network) or direct selection on the basis of the network property.

Other authors have focused their work on the study of network evolution from the kinetic point of view [7,9,15,16]. They obtain mathematical equations from analytical models which defines the kinetic behavior of regulatory network, after that an analysis of network evolution has been performed: convergence to steady states, oscillations, ...

When we want to study evolution in biological systems "in silico", we need to use models which gather the main basis of evolution: mutations, realistic genotype-phenotype mapping, and selection. Moreover if the goal is to study evolution of regulatory networks, we should take into account that mutations don't affect directly the network. Obviously, mutation occurs on the genome, thus modifying indirectly the genetic network. Moreover, the genetic network is not submitted to a direct selection pressure (e.g., for connectivity or specific dynamic properties). The network *contributes* to the cell phenotype. So it is indirectly selected if it produces sufficient effect on phenotype. While many evolutionary studies directly select the individuals for its topological (e.g. small world or scale free) or kinetic properties (e.g. stability or robustness) it is not clear whether such properties have a real impact on the phenotype. Individuals are selected by their phenotype. They are not selected on the basis of the connectivity of their genetic network or their degree of connectivity, for example.

If we want to study evolution in regulatory networks, we need to create an integrated model that includes a realistic enough genotype-phenotype mapping and that will provide us tools to study and analyze the evolution of a genetic networks inside such a system. This biological model should also collects the main biological bases about regulation. It should be stratified on different organization levels from genome to organism's phenotype and it should be compliant with some minimal characteristics of real regulatory networks: different binding levels, auto-regulation, different levels of protein production ... Such a model will not be the model of a specific organism (although our model is mainly inspired from bacterias). On the opposite, the model will gather the general properties of evolution and regulation. Thus, it will hopefully provides us new insights on genetic networks, as well as comparative analysis of genetic networks gave new insights on their general properties.

3 Regulation in Artificial Evolution, the Raevol Model

In order to build such a model we use an integrative evolutionary model previously defined in our team: the Aevol Model [12,13,14].

This model is well suited for our study because it integrates a complex genotype-phenotype mapping including an RNA level between genes and proteins, and different protein concentrations depending on the RNA promoter. Moreover, aevol is able to manage a population of organisms. In this populations individuals undergo mutations and reproduce with different efficiency depending on the metabolic processes performed by their proteins. Thus, the minimal conditions for an evolutionary process to start are already in place in the model.

Aevol has been initially developed to study robustness and evolvability in artificial organisms [?]. However for the sake of simplicity it doesn't include a complete regulation level.

In Aevol model, the artificial genome is made of a variable number of genes separated by non-coding sequences. The genome is organized as a circular double-strand binary string (like in bacterias). Mutations can change the genome structure (by adding/removing new genes or gene order). For each transcribed region, the transcription activity (and thus the protein concentration) is dependent on the promoter sequence. This activity is calculated from the matching level between the RNA promoter and a consensus promoter pattern (the closer the promoter to the consensus, the higher the transcription level). Finally, the organisms phenotype results from the interactions of all the proteins encoded by the genes, modulated by the proteins concentration.

For studying evolution of regulatory networks a regulatory system has been included to Aevol, developing in this way the Regulatory-Aevol Model (Raevol).

In Raevol model, the basal protein's production level is still calculated from the matching level between protein promoter and a consensus promoter but proteins are now able to interact with DNA to regulate the transcription activity. Thus, the core of our model is the protein. In Raevol a protein can have two different roles: it has a regulatory activity, being able to modify other protein's production level, and a metabolic activity. Now, the phenotype results from the interactions of all the proteins that contributes to the metabolism (a protein can have a regulation activity but no metabolic activity), modulated by the regulation network made of all the regulatory proteins (Fig. 3).

3.1 From Genome to Phenotype

The first step is the genome decoding. The genome sequence is parsed to find out promoter position. Once we have localized the promoter regions we look for the Start codon in the transcribed sequence and we extract the binary code of protein. This binary code is translated into a protein sequence by using a functional code. This protein sequence has two tasks: first of all being translated into a metabolic function, contributing to the phenotype, secondly, compute the regulatory activity of the protein (see Fig 3).

In Aevol (and in Raevol), the metabolism is represented as an abstract set of biological functions (see [14] for details). Each proteins contribution to the metabolism is represented as a fuzzy set of triangular shape. It can be described by three parameters: the mean m, the width w and the maximal height H. From gene sequence and using the genetic code, we can determine a Gray binary value associated to the different parameters of this fuzzy set. Protein thus achieve a subset of metabolic processes [13].

When we have translated all proteins in the genome, we have to calculate the interactions between proteins and binding regions to obtain the regulatory graph and calculate the protein production rate (Section 3.2).

Organisms must reach an objective function thanks to an evolutionary algorithm to simulate organisms evolution, so we have to define a phenotype and a fitness function. The **phenotype** is computed as a fuzzy set of processes [13]. If we represent the set of activation functions (i.e., proteins that activate metabolic processes) as A_i and the set of inhibition functions (proteins that inhibit metabolic processes. Represented by a negative h) as I_j the phenotype can be represented as $P = (\cup_i A_i) \cap (\cup_j I_j)$. For this purpose we have used the Lukasiewicz functions, $W(A_1(x), A_2(x)) = \max(0, A_1(x) + A_2(x) - 1)$ for the junction and $W'(A_1(x), A_2(x)) = \min(1, A_1(x) + A_2(x))$ for the intersection.

Fitness: the goal is to compute the individual adaptation to the environment. The environment is represented by an objective function (that represents the best metabolic activity in this specific environment). The objective function is also represented as a set of fuzzy functions. This set can be seen as the ideal set of process E(x) given the environment [13]. We measure the gap g between the objective function and the individual phenotype, the higher the gap the less offspring the organism will have.

3.2 Regulation in genome transcription

To obtain the protein production rate, we should calculate the regulatory capacity of each protein in the individual. By comparing the functional sequence and the binding site (using a regulatory code) we can obtain the protein influence over the regulated protein production (See Fig. 3).

To know exactly how a protein influences the other proteins production, we compare the protein sequence with the DNA regulatory region associated with each promoter. Comparing the functional sequence, and the binding site by using

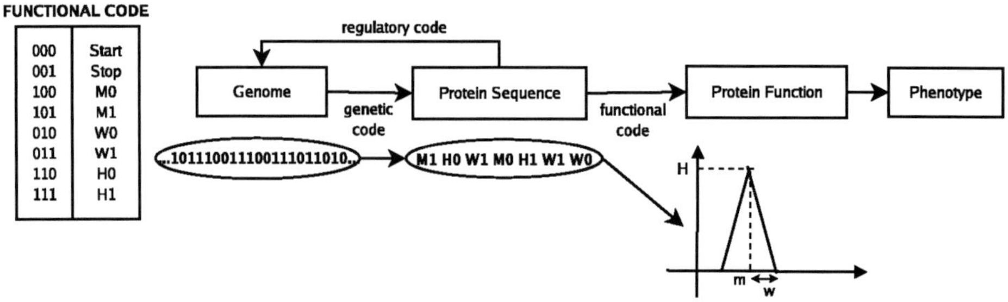

Fig. 3. General schema about steps in DNA-Transcription-Translation and protein behavior levels

a comparison table [3] we can compute the protein affinity with the binding site, thus extracting its influence over the regulated protein concentration.

From each comparison we obtain a value who means how this base can potentially influence the production of regulated protein. Once obtained all the comparison results, mean is calculated. The obtained value enables us to compute how the regulated protein is going to be enhanced or inhibited.

Once we have obtained all binding values, we can build the regulatory graph, and simulate the network behavior. We compute the network concentrations in the cell during twenty time steps. To determine the protein concentration in every time step we use the same equations as in [3]. They provide us a way to reflect the different enhancing/inhibition levels. In these equations protein production is determined by the influence of all proteins over it.

$e_i = \frac{1}{N}\sum_j c_j e^{\beta(u_j^+ - u_{max}^+)}$

$in_i = \frac{1}{N}\sum_j c_j e^{\beta(u_j^- - u_{max}^-)}$

where $i = 1 \dots n$, is the protein number and N the maximum number of proteins. u_{max}^+ and u_{max}^- are the maximum match achievable between protein and binding region, for enhancer and inhibitor proteins respectively. u_j represents the matching value (or binding value) between protein j and the binding region of protein i.

Giving this equations, the produced protein i follows:

$\frac{\partial c_i}{\partial t} = basal\_level(1 + (e_i - in_i) \cdot c_i) - \phi$

Where ϕ is the protein degradation factor.

The initial concentration of each protein is given by its basal level. It corresponds to the matching level between the promoter of the protein and the promoter matching model. For more details see [13].

To compute individual's phenotype fitness, we take the protein concentration levels for each individual to obtain the metabolic functions achieved by the individual (organism phenotype) and we compare them with the objective func-

[3] this table gather the protein codon-DNA base affinity. It has been created following a random normal function and filling half cases with null values

tion. With comparison value we will be able to select the best individuals in population (See Section 3.1).

3.3 Evolutionary Algorithm

As in classical genetic algorithms population evolves by individual mutation. Individual fitness is computed and individuals are selected for reproduction using sampling with replacement. When new individuals are created by parents crossover, they replace parents in population.

Achieved mutations can affect only a few bits (local mutations) or a huge genomic segment. Three types of local mutations can be applied: switching a position (from 0 to 1 or vice-versa) , inserting a few bases (from 1 to 6) in a position, and deleting from 1 to 6 bases. Mutations applied over huge genomic segments can be deletion, duplication , translocation or inversion of long strings [13]. These mutations affect only genome, so they will modify indirectly the protein network topology.

4 Open issues and Future Works

To study network emergence, we have simulated the bacterial behavior in a steady state. Any external protein or influence is going to modify the bacterial metabolic functions. Organisms must reach a set of biological processes (objective function).

We have simulated the evolution in a population of 1000 individuals with a mutation rate of $1e-5$ per base, all aevol parameters are the same as in [?] and regulation parameters are the same as in[16]. These parameters have been chosen to obtain protein concentration stabilization in the last time steps in bacterial life.

After 5000 generation we can see in Figure 4 how individuals in different simulations reach the objective function. In the Raevol model, the lower the gap, the more adapted the individuals. Here we see that the gap value is continuously decreasing, showing that individuals improves their achieving task.

Evolved networks are highly connected (See in Fig. 5, aver. active links) and the network size increases with the number of genes (See in Fig. 5 the average number of genes per genome). However, the links value's histogram (Fig 6) shows that most of these links are close to zero and can thus be considered as inactive links in the network. This is probably due to the stable fitness function we used in this experiment: since the organisms only need to achieve a fixed metabolic function, they evolve a weakly connected network.

In these simulations, individuals have to reach a steady function. They develop mainly negative links (See histogram Fig 6) and they converge to a steady negative value (about -0.025) near to zero (See graphic about mean links in Fig 6).

Although preliminary, these results show that evolution of regulatory networks is not fully random: some features are repeated over the different simula-

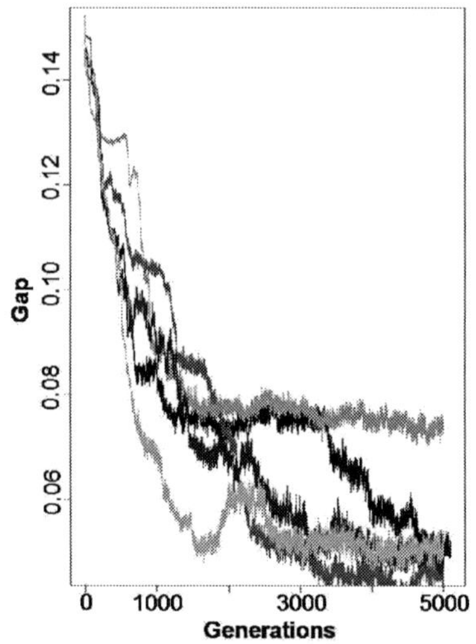

Fig. 4. Mean gap of all individuals in the population over 5000 generations

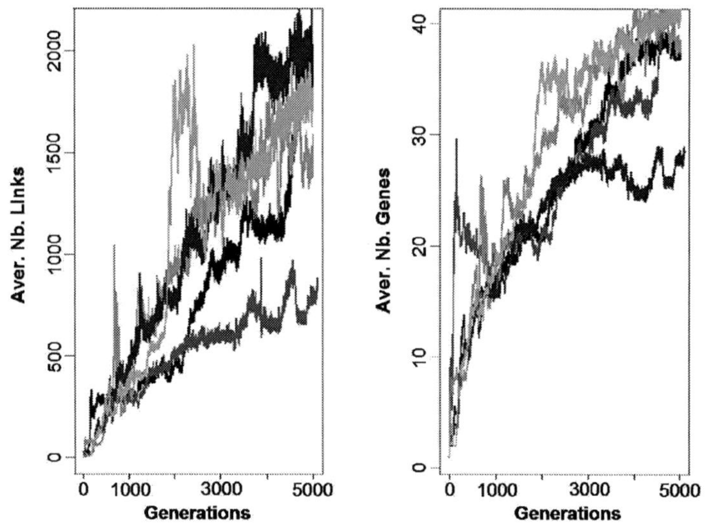

Fig. 5. On the left, evolution of the average number of active links during 5000 generations. On the right, average number of genes per genome. 5 simulations with 5 different seeds during 5000 generations.

tions. This is a very interesting result because networks are not directly selected and it opens a big study field.

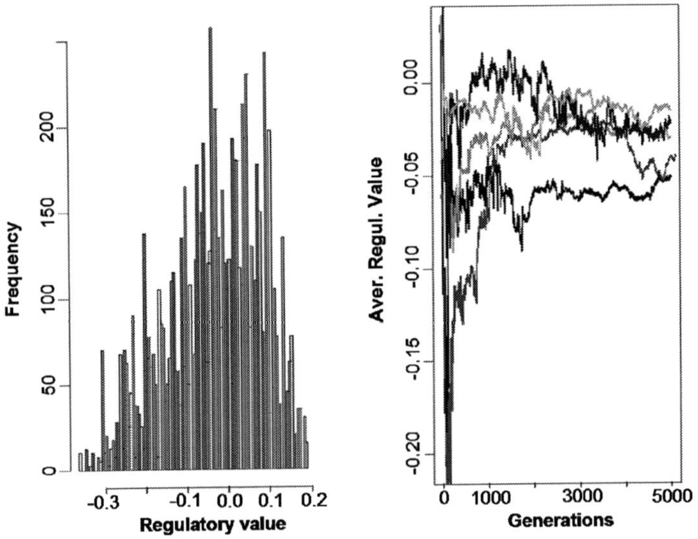

Fig. 6. On the left: Histogram of links values in regulatory network for the best individual at generation 5000. On the right: Evolution of average links values in regulatory network during 5000 generations (for five simulations with five different seeds).

For further experiences, we conceive an other experience: individuals must achieve a set of biological processes, but when an external protein arrive to the nucleus (i.e., a signal), the organisms must change it behavior and stop achieving a subset of biological processes. This external protein should have the inhibitor role in the regulatory network created by the individual. In this case, individuals must adapt themselves not only to develop the first set of processes but also they must adapt their binding sites to the new protein to inhibit the necessary biological functions.

How networks act when individuals receive external events? what's the topology developed by organisms? and how new networks will emerge?, are the next questions we want to answer, thanks to these future experiments.

References

1. M. Aldana, E. Balleza, S. Kauffman, and O. Resendiz. Robustness and evolvability in genetic regulatory networks. *J Theor Biol*, 2006.
2. Wolfgang Banzhaf. Artificial regulatory networks and genetic programming. In Rick L. Riolo and Bill Worzel, editors, *Genetic Programming Theory and Practice*, chapter 4, pages 43–62. Kluwer, 2003.
3. Wolfgang Banzhaf. On evolutionary design, embodiment and artificial regulatory networks. *Embodied Artificial Intelligence*, pages 284–292, 2004. Embodied Artificial Intelligence, F. Iida, R. Pfeifer, L. Steels and Y. Kuniyoshi (Eds.), Springer, Berlin, LNAI 3139, 2004, 284 - 292.

4. Albert-Laszlo Barabasi and Zoltan N. Oltvai. Network biology: Understanding the cell's functional organization. *Nat. Rev. Genet.*, 5:101–113, 2004.
5. Ting Chen, Hongyu L. He, and George M. Church. Modeling gene expression with differential equations. In Stanford University, editor, *PSB*, volume 4, pages 29–40. Stanford University, 1999.
6. Stefano Ciliberti, Olivier C. Martin, and Andreas Wagner. Robustness can evolve gradually in complex regulatory gene networks with varying topology. *PLoS Computational Biology*, 3(2):e15+, February 2007.
7. Hidde de Jong, Michel Page, Celine Hernandez, and Johannes Geiselmann. Qualitative simulation of genetic regulatory networks: Method and application. In *IJCAI*, pages 67–73, 2001.
8. Paul Francois and Vincent Hakim. Design of genetic networks with specified functions by evolution in silico. *Proc. Natl. Acad. Sci. USA*, 101(2):580–585, 2004.
9. Jennifer Hallinan and Janet Willes. Evolving genetic regulatory networks using an artificial genome. *2nd Asia-Pacific Bioinformatics Conference*, 29, 2004.
10. N. Kashtan, S. Itzskovitx, R. Milo, and U. Alon. Efficient sampling algorithm for estimating subgraph concentrations and detecting network motifs. *Oxford Bioinformatics*, 20(11):1746–1758, 2004.
11. Nadav Kashtan and Uri Alon. Spontaneous evolution of modularity and network motifs. *PNAS*, 102(39):13773–13778, September 2005.
12. C. Knibbe. *Structuration des génomes par sĺection indirecte de la variabilité mutationnelle, une approche de modĺisation et de simulation.* PhD thesis, INSA - Lyon, 2006.
13. C. Knibbe, G. Beslon, V. Lefort, F. Chaudier, and J.-M. Fayard. Self-adaptation of genome size in artificial organisms. In M. S. Capcarrere, Alex A. Freitas, Peter J. Bentley, Colin G. Johnson, and Jon Timmis, editors, *Advances in Artificial Life, Proceedings of the 8th European Conference, ECAL 2005*, volume 3630 of *Lecture Notes in Artificial Intelligence*, pages 423–432. Springer, 2005.
14. Carole Knibbe, Olivier Mazet, Fabien Chaudier, Jean-Michel Fayard, and Guillaume Beslon. Evolutionary coupling between the deleteriousness of gene mutations and the amount of non-coding sequences. *Journal of Theoretical Biology*, December 2006.
15. Tommy Krul, Jaap A. Kaandorp, and Joke G. Blom. Modelling developmental regulatory networks.
16. P. Dwight Kuo, Andre Leier, and Wolfgang Banzhaf. Evolving dynamics in an artificial regulatory network model.
17. M. Madan Babu and S. A. Teichmann. Evolution of transcription factors and the gene regulatory network in escherichia coli. *Nucleic Acids Res*, 31(4):1234–1244, February 2003.
18. Orkun S S. Soyer and Sebastian Bonhoeffer. Evolution of complexity in signaling pathways. *Proc Natl Acad Sci U S A*, October 2006.
19. S. A. Teichmann and M. M. Babu. Gene regulatory network growth by duplication. *Nat Genet*, 36(5):492–496, May 2004.
20. Esti Yeger-Lotem, Shmuel Sattath, Nadav Kashtan, Shalev Itzkovitz, Ron Milo, Ron Y. Pinter, Uri Alon, and Hanah Margalit. Network motifs in integrated cellular networks of transcription-regulation and protein-protein interaction. *PNAS*, 101(16):5934–5939, April 2004.

MEDICAL INFORMATICS

Polarimetric images segmentation for lesion detection

S.Mohammed Djaouti, A.Koudache, A.Boudaïeb

University Center Mustapha Stambouli, Mascara, Algeria.
{djouti2007, abd_koudache, boudpg@yahoo.fr}

Abstract. We present in this article a Mueller polarimeter for the assessment of superficial skin lesions. The system is capable of capturing polarized light images non-invasively. There are two objectives in this paper. The first is to describe how to obtain a polarimetric image through a Mueller imaging system. The second objective is using the Markov random field model to segment theses polarimetric images. We are choosing the ICM algorithm for the optimisation of probability because it is speed and can converge in few numbers of iterations. The use of the Markovian model gives efficient results separating normal and pathologic skin (detection of lesions).

Keywords: Polarimetric images, Mueller matrix, segmentation, Markovian model, ICM.

1 Introduction

In recent years, several studies have indicated that polarized light could provide non-invasive diagnosis of skin pathologies, such as malignant pigmented and unpigmented skin sites, irradiated skin, or mucosa.

When a light ray strikes skin, a fraction is reflected by the surface of the skin, while the rest is propagated into the tissues, where it is either absorbed or scattered by hemoglobin, melanin, or cells. Each of these interactions contributes to a change in polarization state. Polarized light reflected or scattered by the air-tissue interface, for example, can be compared to light scattered by an irregular surface and modeled by a facet scattering model [3]. Consequently, the top surface contributes mostly to the polarized part of the backscattered light, especially near the specular direction. Light that propagate deeply into the tissue is usually multiply scattered before escaping and is nearly fully depolarized. Light that is weakly scattered by superficial tissues is expected to partially maintain its initial polarization state. Therefore, deep subsurface and superficial tissue imaging may be possible using polarization imaging.

Polarized light sources provide illumination in and out of the plane of viewing at three different wavelengths and sixteen different directions. A Stokes vector imaging system is used to detect the light and to separate the polarized component from the unpolarized component and to further decompose the polarized component into two components.

CP940, *CompLife 2007, 3rd International Symposium on Computational Life Science*
edited by A. Siebes, M. Berthold, R. Glen, and A. Feelders

Polarization-encoded imaging consists of the distributed measurements of polarization parameters for each pixel of an image. We address clustering of multidimensional polarization-encoded images. The spatial coherence of polarization information is considered. Two methods of analysis are proposed: *polarization contrast enhancement* and a more-sophisticated *image-processing algorithm based on a Markovian model.* The proposed algorithms are applied and validated with different Mueller images acquired by a Mueller polarimetric imaging system.

This paper is organized as follows: in the section 2, we present a background of the physical analysis of Mueller matrix with a brief description of the procedure of obtaining of the Mueller matrices. Section 3 describes the process of segmentation of the polarimetric images highlighting different results obtained after application and validation of different algorithms then we conclude by a conclusion.

2 Physical analysis of Mueller matrices

Mathematically, Mueller matrices act as linear operators on the incoming Stokes vectors (S') yielding the outgoing (S) Stokes vectors. It relates the Stokes vector of the incoming waves linearly to the Stokes vector of the outgoing ones [2]. It characterizes the transfer of energetic and polarizing properties of the system to the beam. Formally, this can be written as

$$S' = M.S, \tag{1}$$

$$S = \begin{bmatrix} I \\ Q \\ U \\ V \end{bmatrix}, \tag{2}$$

$$[M] = \begin{bmatrix} m_{00} & m_{01} & m_{02} & m_{03} \\ m_{10} & m_{11} & m_{12} & m_{13} \\ m_{20} & m_{21} & m_{22} & m_{23} \\ m_{30} & m_{31} & m_{32} & m_{33} \end{bmatrix} \tag{3}$$

The main interest of this approach lies in the addition theorem of Stokes vectors and the optical equivalence principle, allowing decomposition of a general matrix as the incoherent sum of matrices representing “simpler” processes [2]. Furthermore, Mueller matrices are able to represent depolarizing systems in contrast with Jones vector-matrix formalism. Although it is always possible to associate any (non-depolarizing) system’s Jones matrix to its Mueller matrix, the reverse is not always true.

2.1 Mueller matrix optical backscattering image system setup

The Mueller matrix was measured with an optical polarimeter depicted in Fig. 1 [5], [6]. The input polarization encoding systems composed of a vertical (x-axis) linear polarizer, P_1, followed by a quarter-wave plate whose fast axis makes an angle θ with the x-axis. The output polarization decoding system consists of a quarter-waveplate, whose fast axis makes an angle θ' with respect to the reference x-axis; it is followed by a linear horizontal polarizer, P_2. P_1 and P_2 were initially crossed in order to obtain a null intensity. The Stokes vector S emerging from P_2 can be expressed by the following matrix product:

$$\vec{S}' = [P'][L'][M][L][P]\vec{S} \tag{4}$$

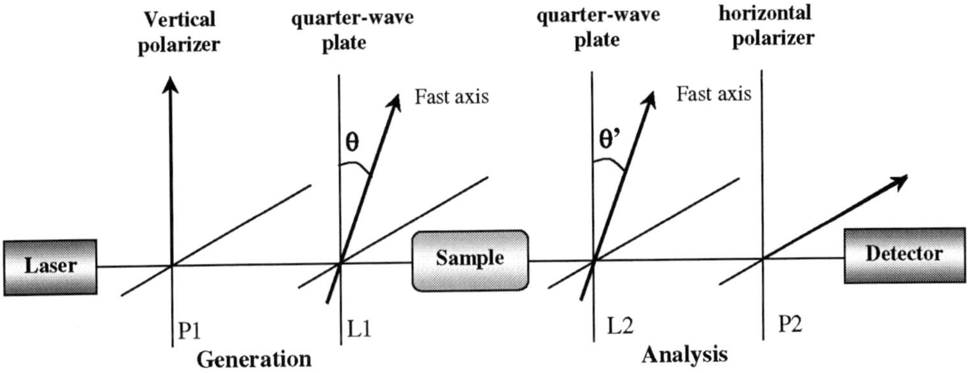

Fig. 1. Experimental setup.

where $[P_2]$, $[L_2]$, $[L_1]$ and $[P_1]$ are the Mueller matrices of each of the devices constituting the polarimeter, [M] is the unknown Mueller matrix and S is the Stokes vector of the incident light. The measurable intensity is enclosed in the first term of the output Stokes vector.

$$[P] = \frac{1}{2}\begin{bmatrix} 1 & 1 & 0 & 0 \\ 1 & 1 & 0 & 0 \\ 0 & 0 & 0 & 0 \\ 0 & 0 & 0 & 0 \end{bmatrix}, \text{ vertical polarizer} \tag{5}$$

$$[P'] = \frac{1}{2}\begin{bmatrix} 1 & -1 & 0 & 0 \\ -1 & 1 & 0 & 0 \\ 0 & 0 & 0 & 0 \\ 0 & 0 & 0 & 0 \end{bmatrix}, \text{ horizontal polarizer} \tag{6}$$

$$[L]=\begin{bmatrix}1 & 0 & 0 & 0\\ 0 & C^2 & CS & -S\\ 0 & CS & S^2 & C\\ 0 & S & -C & 0\end{bmatrix} \quad \begin{matrix} C=\cos(2\theta) \\ S=\sin(2\theta)\end{matrix}, \text{ quarter-wave plate} \tag{7}$$

One needs only 16 equations to obtain the Mueller matrix coefficients. To minimize errors we used an over determined system of 64 equations corresponding to 64 combinations of the angles (θ, θ'). In our experiments, these angles were multiples of 22.5° to describe specific test positions.
Theoretically, the Mueller matrix with no sample should be the identity matrix. But, because of retardance errors on the two quarter-wave plates, experiments have shown that, on the one hand, the elements m_{01} and m_{10} of the 4 × 4 Mueller matrix calculated from experimental data were not equal to zero and, on the other hand, m_{11} and m_{22} were not equal to 1. In addition, simulations proved that the correlation between the retardance-error-caused deviations on m_{01} and m_{10} is only very weak when the retardance errors are small (less than 5°): indeed, any retardance error on the input plate will result in an error on m_{01}, but this is not true on m_{10}. On the other hand, any error on the output plate will cause an error on m_{10}, but none on m_{01}. Further, we will calibrate our polarimeter in order to minimise theses errors (random errors or systematic errors).

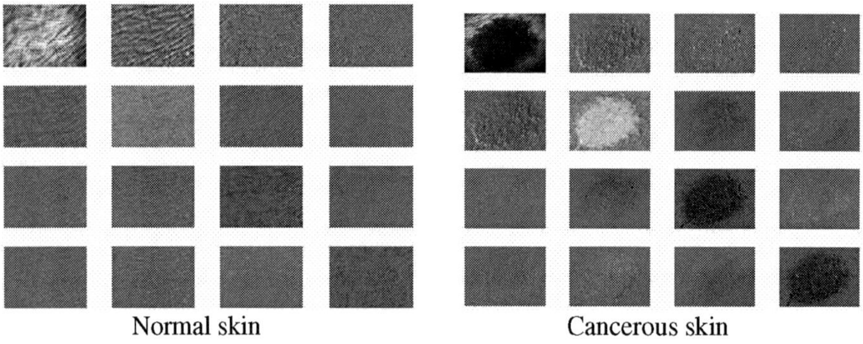

Fig. 2. The results of the Mueller imaging system.

After getting the Mueller images, we pass to the second step witch consists of the segmentation of theses images.

3 Polarimetric images segmentation

Image segmentation remains a difficult issue, particularly when what is unsupervised. As the state of instrumentation evolves to produce increasingly finer resolution in spectral, spatial, and temporal data, increasingly more sophisticated techniques are required for handling these data properly. Statistical approaches have proved successful as robust methods of analysis and segmentation. In the context of multihyperspectral images or multimodal–multiresolution images, handling correlated observed data requires a well-designed modelling framework. In this way, the past decade has provided a rich variety of model proposals (Markov Model, polar decomposition or dimensionality reduction) [7].

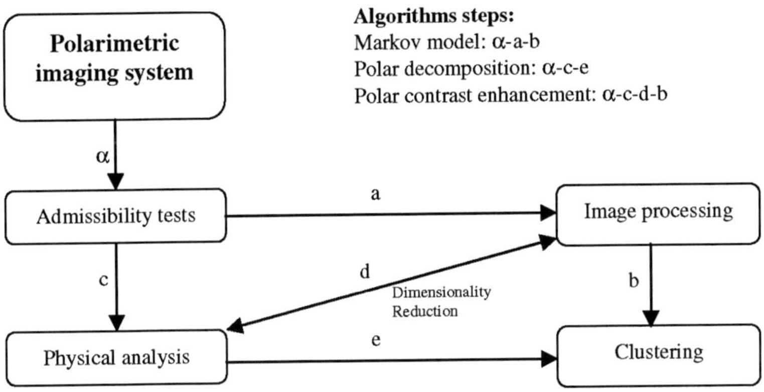

Fig. 3. Steps in the three algorithms for clustering Mueller images [1]. The Markov-model-based algorithm directly processes 16 Mueller channels (steps α–a–b). The polar decomposition algorithm is mainly a physical processing of polarization data (steps α–c–e). The polarization-contrast enhancement algorithm uses polarization-based information to reduce the data's dimensionality and to provide a more suitable image for classic image-processing algorithms.

A Bayesian scheme based on a Markov model is attractive when one is dealing with large amounts of data [7]. In the context of multimodal polarimetric images, we propose to adopt a Markovian model that exhibits a good ability to learn the required model parameters in an efficient and fast manner.

3.1 Markovian Model and Bayesian Inference

Our motivations for using Markov Model are to provide fast computations and efficient structures to analyze parameter data cubes. Indeed, different communities needs tools to interpret large data corresponding to multispectral observations (e.g., astronomical images), parameter observations (e.g., polarimetric imaging) or volume information (e.g., X cubes for medical applications), etc.

The segmentation problem is to estimate the unobserved realization $X = x$ from the observed realization on $Y = y$, where: $y=y_s$, $s \in S$ and S is the set of voxel positions in the data cube to be segmented. X_s takes their values in $\Omega = (\omega_1, \omega_2, \omega_3, \ldots, \omega_K)$ (k is the number of classes).

The segmentation step is thus of great importance to cluster the pixels into sets of similar spectral, statistical or geometric features, in order to be able to extract their characteristics with further discriminate analysis (classification step) [1],[4]. In the context of unsupervised Bayesian segmentation, we need to estimate all the parameters defining the joint distribution P(X,Y) between the label field X that have to be estimated and the noisy observed field Y. In other words, PX is the marginal distribution of X, and P(Y/X) is the distribution of Y conditional to X. P(Y/X) is called the data driven term and $P(Y/x) = \prod_{s \in S} P(Y_s / X_s)$ where $P(Y_s / X_s)$ follows a Gaussian distribution. The Gaussian parameters for $P(Y_s / X_s)$ are estimated by Expectation Maximization (EM, SEM, ICE ...) processes [10]. Under Markovian assumption, the X distribution takes into account neighbourhood labels (spatially or in scale) on each site in order to provide a form of context handling in the labeling process. This a priori favors homogeneity in the labeling process.

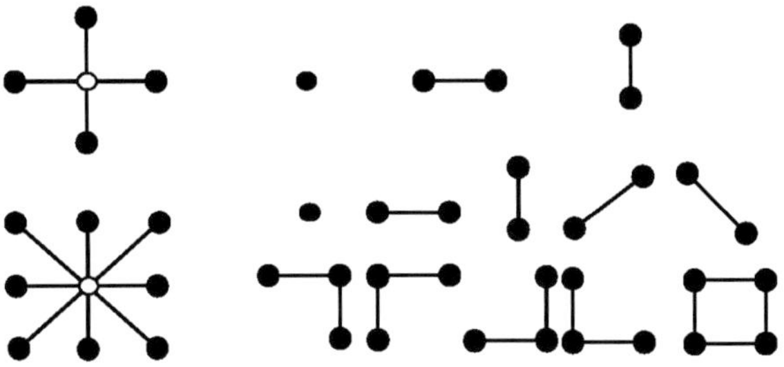

Fig. 4. Neighboring system [9].

$$N_s = \{t\} \qquad with \begin{cases} s \notin N_s \\ t \in N_s \Rightarrow s \in N_t \end{cases} \tag{8}$$

Let us assume that the X distribution is Markovian, i.e.,

$$P(X_s = x_s / x^s) = P(X_s = x_s / x_t, t \in N_s) \text{ with } x^s = (x_s)_{t \neq s} \tag{9}$$

Then X follows a Gibbs distribution with respect to a neighbouring system (a Gibbs random field governing the grouping of image pixels into regions) , for example the set of the four nearest neighbours of $s \in S$. Denoting by C the set of cliques (a clique being either a singleton or a set of sites mutually neighbours), the X distribution is:

$$P(x) = \frac{1}{Z} \exp(-U_2(x)) \tag{10}$$

where Z is a normalization coefficient given by $Z = \sum_{x \in \Omega} \exp(-U_2(x))$ and $U_2(x)$ a potential function (sum of energy of each clique).

Getting the optimal class needs necessary to maximize the global joint distribution P(X,Y) or minimize the global energy:

$$\hat{x} = \arg\min_{x \in \Omega} \quad (U_1(y/x) + U_2(x)) = \arg\max_{x \in \Omega} \quad P(x/y) \tag{11}$$

With $U_1(y/x) = -\ln P(y_s/k)$.

For the optimisation of posterior probability, we have used the ICM algorithm that affect the voxel to the class having the minimal energy (compared to the simulated annealing and MPM, it is speed and can converge in few iterations to the global minima due to the initialisation of the algorithm).

3.2 The iterated conditional modes: ICM algorithm

The defined spatial segmentation is based on the ICM algorithm introduced by Besag. This algorithm is used for image reconstruction when local true image characteristics are modeled under Markovian principles. Specifically, this approach can be used collaboratively with the Potts/Ising model. This iterative algorithm commences using an initial estimate of the true scene x, referred to as $\hat{x}$, and proceeds to compute estimates for the parameters associated with $P(Y_s / X_s)$. Considering the X values are latent and the original image voxels Y_s can be observed, an assumption is made that the conditional density model, linking the observed and latent spaces $f(Y_s / X_s = j)$, follows a Gaussian distribution with mean m_j and variance σ^2.

The algorithm proceeds as follows:

i) Initialisation of the ICM algorithm (with parameters estimation of each class).

ii) From the initial configuration, theses steps are realised for each pixel:

a) For each class :

1. Calculation of U_1;
2. Calculation of U_2;
3. Calculation of the global energy $U = U_1 + U_2$.

b) Search the class of minimal energy.

iii) Affect to the current pixel the class returned from b).

iv) ICM concludes either on convergence of $\hat{x}$ or when a predetermined number of iterations have been reached. In this instance, the maximum number of iterations is set to 20; however, in other studies only ten iterations were utilized [7], [8].

4 Results and discussion

To attain our objective, the following steps are respected:

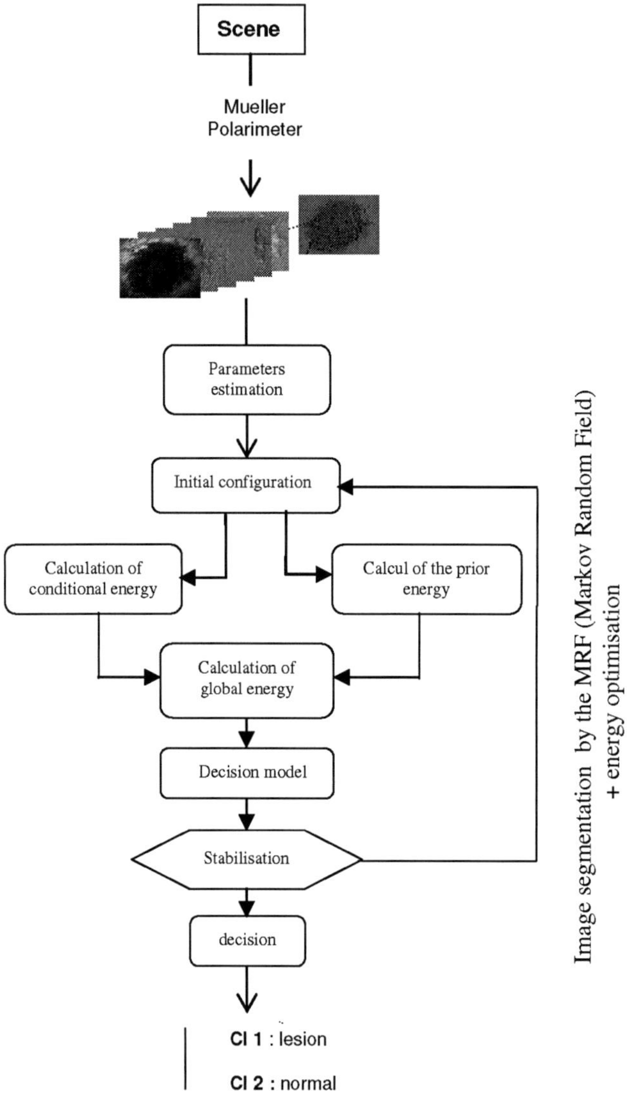

Fig. 5. The gait of segmentation.

Here we present some sample results for the image models used and segmentations obtained using the adaptive process described in the previous section. In each of the segmentations depicted, the algorithm incorporated parameter estimation as described above, but for the sake of brevity we do not include the specific details and the quantitative data associated with the convergence of parameter estimates.

Fig. 6. The 16 images after segmentation (using ICM).

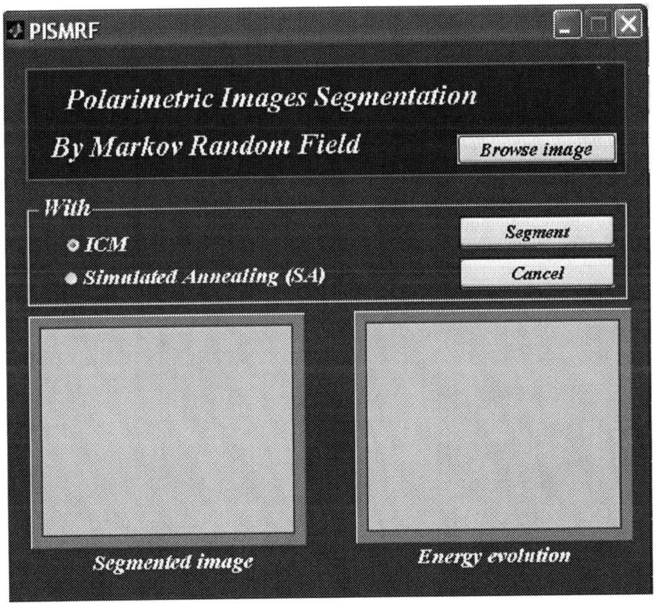

Fig. 7. The application interface.

The number of image (eg. 11) is simulated as the Mueller matrix.

Image 00

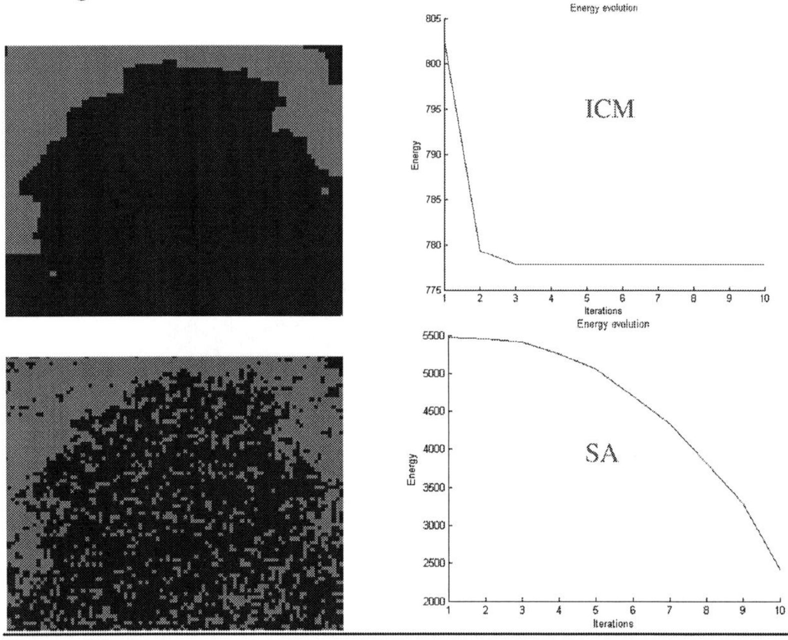

Image 33

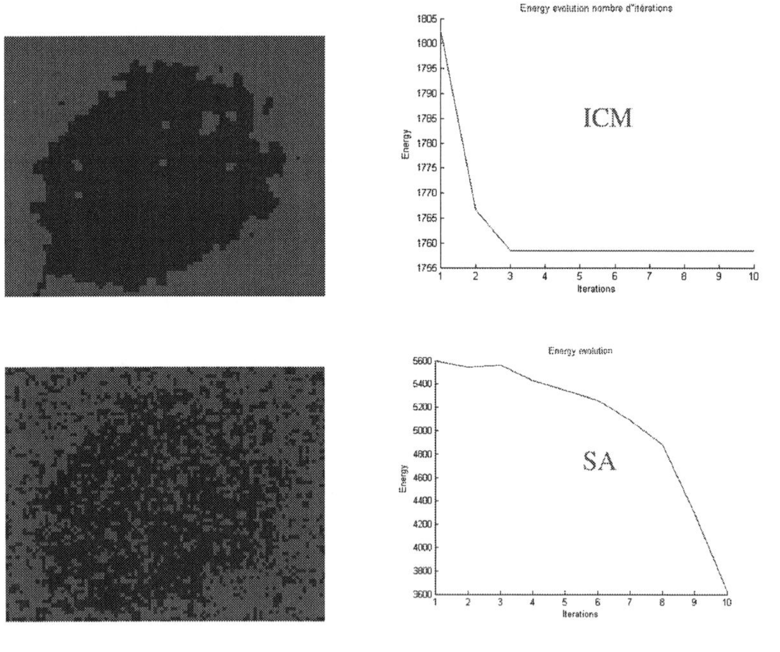

Fig. 8. Results of segmentation of two images with the convergence of the ICM algorithm.

Fig. 6 shows the result of application of our strategy (the interface of the application is given on Fig. 7) with the use of one of the algorithms: either ICM or SA. We can see that diagonal images could give best results. Each image of the collection (Fig.6) cooperates in order to accomplish the information given by the image witch proof the importance of each image.

Fig. 8 demonstrates the result of segmentation of two images with the use of two optimisation algorithms (the number of iterations is fixed at 10). We can see very well the efficient of the Markovian model combined to the ICM to separate normal skin from the pathologic one in few iterations (it is linked to the initialisation of the ICM algorithm that converge to the first minimum global) indeed, compared to the ICM, with the SA algorithm, the number of iterations is augmented to more than four iterations (SA can give good results if we augment the number of iterations) with the non convergence of the energy.

We propose as future work the fusion of the 16 images to obtain a multivariate image to segment.

5 Conclusion

Polarimetric imagery can be used effectively to characterize objects even under conditions of low radiance. It is particularly useful in determining the geometric form of a shape and composition of targets. In this study we have signalled the usefulness and the importance of coupling polarization information with image-processing techniques. Considering the bidimensional consistency of polarization information is shown to be very efficient. However, a Bayesian scheme based on a Markov model is attractive with large amounts of data because it takes in consideration the notion of spatial connectivity.

6 References

1. Zallat, J., Collet, C. et Takakura, Y.: Clustering of polarization-encoded images. Applied Optics, vol. 43, N° 2, 2004.
2. Vézien, C., Clairac, B., Fontas, B. et Guyot, S.: Conceiving an Easy-to-Understand and Automated Polarimeter: Application to a Study of Collagen. IEEE Transactions On Education, Vol 48, N°4, 2005.
3. Terrier, P., Devlaminck, V.: Système polarimétrique pour l'analyse d'images. Traitement du Signal 2000, Vol 17, n°5/6 (2000) 479-490.
4. El Ghouat, M. A. : Classification markovienne pyramidale : Adaptation de l'algorithme ICM aux images de télédétection. Thesis, Sherbrooke university –Canada, 1997.
5. Boulvert, A.: Analyse de milieux fortement diffusants par polarimétrie de Mueller et méthodes optiques cohérentes. Application à l'étude du syndrome cutané d'irradiation aiguë. Thesis, Bretagne Occidentale university 2006.
6. Boulbry B.: « Mise en place d'un polarimètre de Mueller achromatique dans le domaine du visible en régime laser impulsionnel ». Thesis, Bretagne occidentale university, 2002.

7. D. Stanford, "Fast automatic unsupervised image segmentation and curve detection in spatial point patterns," Ph.D. thesis, University of Washington, 1999.
8. Zaidi H., Abbes A. and Montgomery D. W., Fully automated segmentation of oncological PET volumes using a combined multiscale and statistical model, Med. Phys. Vol 34, N°2, 2007.
9. A.Zuerguerras, A.Boudaieb, Segmentation non supervisée d'images SAR par application de l'algorithme EM, Algerian Journal of Technology, vol.14, n°1, 1999.
10. Benboudjema D., Champs de Markov triplets et segmentation bayesienne non supervisée d'images. Thesis, Troyes university of Technology, 2005

Studying channelopathies at the functional level using a system identification approach

A. Aldo Faisal
aaf23@cam.ac.uk
[1]Computational & Biological Learning Lab, Dept. of Engineering, Cambridge University, CB2 1PZ Cambridge, UK

Abstract. The electrical activity of our brain's neurons is controlled by voltage-gated ion channels. Mutations in these ion channels have been recently associated with clinical conditions, so called channelopathies. The involved ion channels have been well characterised at a molecular and biophysical level. However, the impact of these mutations on neuron function have been only rudimentary studied. It remains unclear how operation and performance (in terms of input-output characteristics and reliability) are affected. Here, I show how system identification techniques provide neuronal performance measures which allow to quantitatively asses the impact of channelopathies by comparing whole cell input-output relationships. I illustrate the feasibility of this approach by comparing the effects on neuronal signalling of two human sodium channel mutations (NaV 1.1 W1204R, R1648H), linked to generalized epilepsy with febrile seizures, to the wild-type NaV 1.1 channel.

1 Introduction

Voltage-gated ion channels confer excitability to neurons, allowing them to generate the fundamental electrical signal in our brains, the action potential. The direct implication of mutations in these ion channels in diseases recently spawned the field of 'channelopathies' [1,2,3,4,5,6]. To understand the symptomatic basis of such ion channel-linked genetic disorders, it is not only crucial to understand how a gene defect will alter the molecular process of channel gating, but also how this alteration in channel kinetics will affect whole cell and network dynamics. The experimental approach to this question is the use of transgenic animals, which is becoming established [7].

In contrast, here, I follow another route: through theoretical and simulation techniques one can build virtual "transgenic" neurons and networks, thereof, that provide us with unique investigative tools [8]. This enables us to look at the system's level impact of experimentally obtained molecular data (here, ion channel gating kinetics). Such data can be gathered much easier and economically from isolated gene expression systems (*Xenopus* oocytes) than whole transgenic animals. This bottom-up approach allows to link molecular data (e.g. ion channel kinetics) with the resulting pathophysiology, which will be otherwise difficult to uncover using a top-down approach involving transgenic animals as one cannot

CP940, *CompLife 2007, 3rd International Symposium on Computational Life Science*
edited by A. Siebes, M. Berthold, R. Glen, and A. Feelders

control the individual system parameters as easily. Furthermore, recent progress on the role of ion channel properties in the architecture of the brain as a whole [9,10,11] prompt a shift from the detailed single ion channel biophysics and protein structure studies towards the interplay between the characteristics of the channel protein and their effects at the whole cell and network level. This motivates the following novel approach to channelopathies: The gating-kinetics of ion channels affect the activity ("spike pattern") by which a neuron responds to patterns of synaptic input, and thus the neuron's overall input-output relationships. The way the input-output relationship depends on the detailed ion channel kinetics is complex, due to the non-linear nature of the action potential mechanism. Therefore, input-output relations, which characterize system-level behaviour, cannot be captured by straightforward experimental approaches (e.g. measuring the average spike rate in response to constant stimulation [8]). Here, I will demonstrate that by using an appropriate characterization method one can render the effects of channelopathies on a neuron's input-output relationship detectable by computational methods – not only between different types of channels (e.g. [11]) but also for point-mutations of ion channel proteins.

How can we describe the input-output relationship of a system to predict the response of the system to arbitrary inputs and to assess the effects of channel mutations? I combine two approaches: The first approach is the classic method of analysing the individual ion channel kinetics and construct from them a set of coupled differential equations (or other) equations and determine the parameters of these coefficients that best predict system performance, as applied since Hodgkin and Huxley modelled squid axon [12] and recently also in channelopathies [8]. This allows very accurate modelling of neurons, reproducing and predicting physiological properties to an exquisite degree. However, an intuitive understanding of how individual ion channels properties contribute to overall neuronal behaviour is lost. Therefore, I contrast (and will then combine) this first, synthetic approach with a second one: the non-parametric "black box" approach of system identification [13,14]. Here, the aim is to model a system's input-output characteristic without specifying the internal structure or mechanisms present. Mathematically speaking, the system's functional $F, y(t) = F[x(t)]$ is modelled, where $x(t)$ and $y(t)$ are the input and output of the system. Using appropriate inputs $x(t)$ and observing the response $y(t)$ the system functional F can be estimated. For inputs $x(t)$ that are (band-limited) white noise the system's functional can be decomposed into a linear combination of individual Wiener kernels of increasing complexity, which combined together have exactly the same transfer characteristics as the system under study. In the case of neuronal systems where the input is a white noise sensory or current stimulus and the output is the train of action potentials (spike train) we can truncate the Kernel expansion after relatively few Kernels (here, two), as these will typically carry the majority of the systems input-output characteristics [15,14,16].

To illustrate the feasibility of this approach for channelopathies I look at the effects of mutations in a sodium (Na^+) channel associated with clinical conditions: The human voltage-gated Na^+channel NaV1.1 (SCN1A_HUMAN,

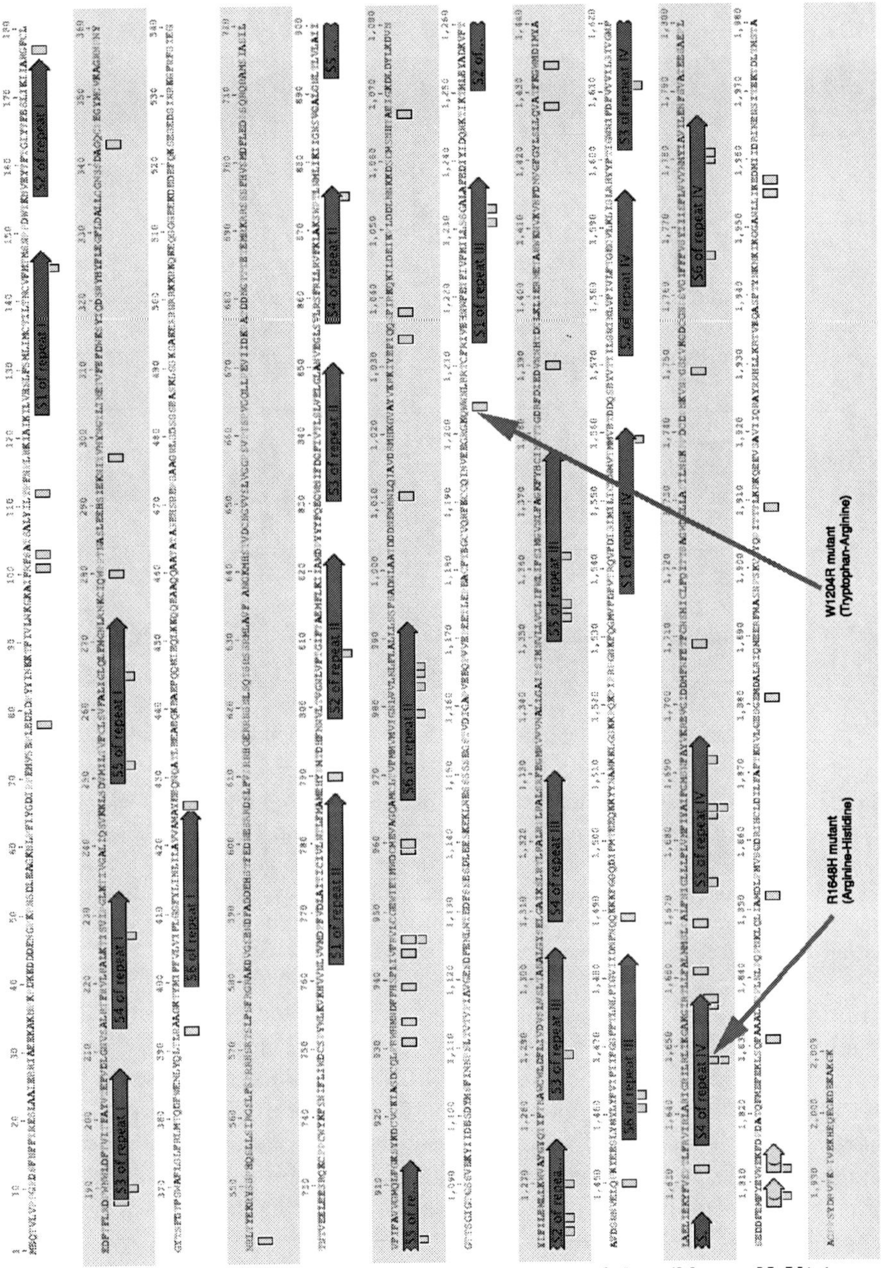

Fig. 1. (vertically oriented) Annotated protein sequence of the wild-type NaV1.1 protein (SCN1A_HUMAN), a human voltage-gated sodium ion channel (Accession number P35498, Uniprot database). Purple arrows indicate transmembrane regions (S1-S6) of the four channel alpha subunits (I-IV). Note, the S4 region is associated with voltage-sensing. Yellow boxes (and arrows) indicate presently known sequence variants or mutants. Red arrows mark the position of the two point-mutations R1648H and W1204R discussed in the text.

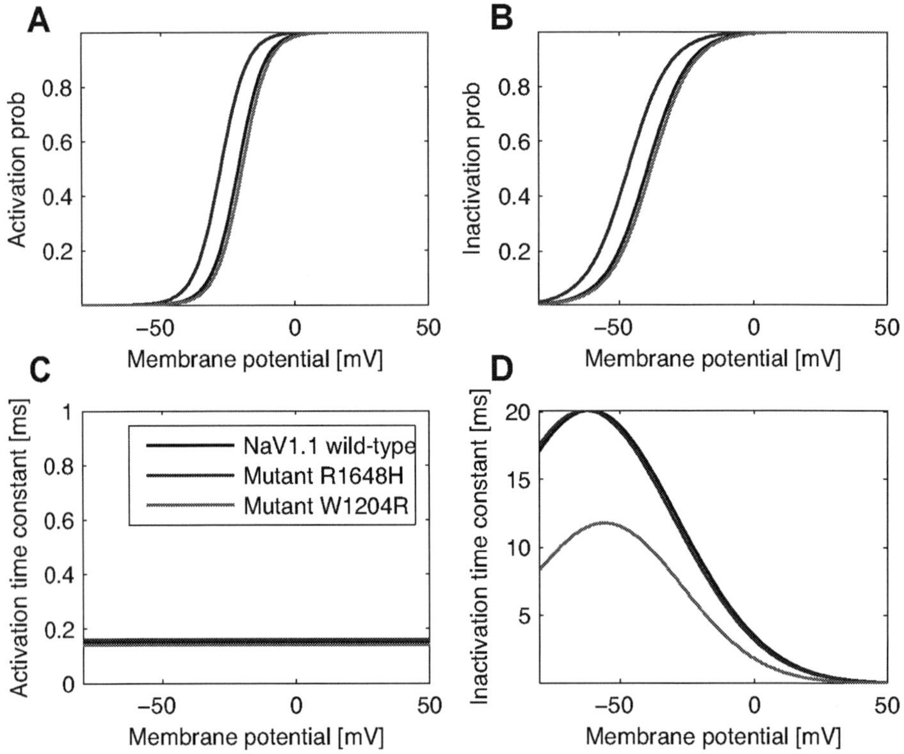

Fig. 2. Overview of channel kinetics for the mammalian sodium channel NaV1.1 (Wild-type, black) and two known mutants W1204R (blue) and R1648H (red), both found in humans. (A) Voltage-dependent steady-state activation probability $m_{\infty}(V)$. (B) Voltage-dependent steady-state inactivation probability $h_{\infty}(V)$. (C) Channel activation time constant $\tau_m(V)$. (D) Channel inactivation time constant $\tau_h(V)$. Note, typical resting potential for associated neurons is −60 mV.

UniprotKB/Swiss-Prot accession number P35498, see sequence in Fig. 1) is mainly found in the soma of neurons in the cortex, hippocampus, cerebellum, spinal cord, brain stem [2]. Mutations of this ion channel are known to cause inheritable forms of epilepsy. The mutated and wild-type channels were previously characterized, isolated and then expressed in *Xenopus* oocytes, allowing their channel kinetics to be determined in detail [7,8] using *in vitro* voltage-clamp or patch-clamp techniques. Fig. 2 illustrates the kinetic characteristics of the three channels.

The first mutation, Swiss-Prot variant VAR_010111 in P35498, or short "R1648H" is a genetic defect, where the amino acid substitution of Arginine by Histidine at position 1648 in the NaV1.1 channel protein has been shown to cause familial epilepsy [17]. The mutation affects channel kinetics in a voltage-dependent manner as it is located in the voltage sensor region of the trans-membrane segment of the protein. Correspondingly, the R1648H mutant has activation and inactivation characteristics (blue curve in Fig. 2.A,B) that are shifted lower by a few millivolts in comparison to the wild-type channel (black curve).

The second mutation, Swiss-Prot variant VAR_014270 in P35498, or short "W1204R" was identified in a large screen of human families symptomatic for generalized epilepsy with febrile seizures plus "GEFS+" [18]: it consists of a point mutation at position 1204 which substituted Tryptophan for Arginine. The W1204R mutation is located in an evolutionary highly conserved region of the fourth subunit: between the second and third transmembrane domains. This region could be implicated in sodium channel inactivation and correspondingly, as the W1204R channel's kinetics (red curve) has a factor two lower inactivation time constant (black curve).

How do these point-wise differences at the genetic level affect system-level behaviour?

2 Methods

I obtained a neuron's response to dynamical stimulation with naturalistic fluctuating current stimuli (as experienced *in vivo*) by using detailed simulations of neurons. Using simulations instead of experiments allowed me to control all parameters, and so concentrate on differences induced by mutations and not by other sources of biological variability. The fluctuating stimulus was a white noise process with zero mean and 0.07 nA standard deviation, producing a stimulus that resembles macroscopically the synaptic input driving many neurons in the cortex [19]. The stimulation lasted 3000 s in each trial.

2.1 Simulation of neuron behaviour

The membrane response $V(t)$ of a model neuron is the solution of an integral-differential equation which is not analytically solvable, (without approximating away the specific kinetic properties of ion channels [20]). Therefore, I used my purposely developed software, the Modigliani stochastic simulator [21,10] (see http://www.modigliani.co.uk) to simulate 3 types of neurons which differed only in the respective type of Na^+channel used (NaV1.1 wild-type, R1648H, W1204R). Stimulation inputs, as well as all other experimental protocol conditions where identical.

Neurons had a cylindrical shape, with diameter and length of 10 μm, modelled using 5 equal-sized membrane compartments and thus similar to cortical cell bodies in shape and membrane properties. The spatial characteristic allowed membrane potential differences between stimulus site (one end of the cylinder) and the site where membrane potential was recorded (at the other end of cylinder). In contrast to other simulation studies using single compartments, this decoupled the measured cell response from stimulus artifacts to a certain extent. The neuron's membrane leak conductance was $g_L = 0.1\ mS/cm^-2$, membrane capacitance was $C_m = 1\ \mu F/cm^2$, and cytoplasmic resistance $R_a = 70\ \Omega cm$.

The membrane contained one of the three NaV1.1 sodium channels types discussed above (see [8] for the tabled kinetic rate equations), channel densities were 100 μm^2 and single channel conductance was $\gamma_{Na} = 20\ pS$ in all three cases. The membrane contained K^+channels with SGAK kinetics (Hodgkin-Huxley's delayed rectifier K^+channels, [12]). K^+Channel density was 28 channels per μm^2 and single channel conductance $\gamma_{Na} = 20$ pS. Resting potential was -60 mV with a leak reversal potential of $E_L = 19.573$ mV for all three Na^+channel types used. Channel kinetics were adjusted to a common temperature of 6.3°C and simulated using Modigliani's built-in Crank-Nicholson PDE solver (simulation time step size was 0.001 ms). Note, that Na^+conductance and passive electrical properties (such as resting potential) where the same for all 3 types of neurons studied.

2.2 System identification

I will next only briefly state the system-identification approach for neurons, because the details of Wiener kernel analysis and their adaptation to physiological/neuronal systems have been previously exquisitely discussed (see [14] for derivation and [15,22,23] for adaptation details).

The idea of system identification is to derive from a known input $x(t)$ and output $y(t)$ the internal properties F of a system. In our case, the input $x(t)$ is the current stimulus and the output $y(t)$ consists of the electrical impulses ("spike train") representing a "1" whenever an action potential occurred (defined as an upward crossing of membrane potential 50 mV above resting potential), and "0" otherwise. The system is the whole neuron.

The zeroth-order Wiener kernel h_0 represents the average output of of the system, here the average firing rate of the neuron.

$$h_0 =< y(t) >_t \tag{1}$$

where $< \cdots >_t$ denotes time averaging.

The first-order Wiener kernel is the average value of the stimulus current $x(t)$ at a time τ before the occurrence of a spike. This value is normalised over the total stimulation time T and the power of the current stimulus σ_x^2.

$$h_1(\tau) = \frac{1}{\sigma_x^2} \int_o^T y(t)\ x(t-\tau)dt \tag{2}$$

Thus, h_1 measures the simple correlation between stimulus and response. For linear systems, h_1 is the impulse response, however for nonlinear systems such as spiking neurons, there are contributions from higher orders of stimulus correlation. Consequently, the second-order Wiener kernel h_2 is obtained by cross-correlation between the input stimulus $x(t)$ at two different instants τ_1 and τ_2 and the system's spike train response $y(t) - h_0$.

$$h_2(\tau_1, \tau_2) = \frac{1}{2\sigma_x^2} \int_0^T (y(t) - h_0)x(t-\tau_1)x(t-\tau_2)dt \tag{3}$$

The h_2 kernel, measures the interaction between two parts of an input stimulus trace that generated a spike. Thus, the second order kernel captures the response deviation from the linear superposition of the responses to each input $x(t - \tau_1)$ and $x(t - \tau_2)$.

For practical purposes, the kernels h_0, h_1, h_2 are calculated in discrete time (sampled at 20 kHz), thus the integrals become sums over discretised time. The h_1 kernel is a vector of length l and the h_2 kernel is a symmetric matrix of size $l \times l$. I chose $l = 250$, which corresponds to the first 25 ms prior to a spike, each of the l dimensions corresponds to a time bin of width 50 μs$= \frac{1}{20\ kHz}$. The 25 ms duration was always longer then the significant non-zero components of the first- and second-order kernels, so longer time correlations could be ignored.

Next, I computed the extent to which the stimulus, $x(t)$, could be reconstructed from the recorded spike trains $y(t)$ [22]. To that aim the true stimulus has to be compared to it's estimated stimulus $\hat{x}(t)$:

This can be obtained by convolving the input $x(t)$ (less its constant average, the zero-th order kernel h_0) with the first order kernel h_1:

$$\hat{x}(t) = \int h_1(t - \tau)(y(\tau) - h_o)d\tau \tag{4}$$

Strictly speaking, this equation holds only if the spike train x(t) is Poisson-like, which is typically satisfied at low spike rates. In this case individual spikes are independent of each other such that they dot not interact via their refractory periods [23]. This is the case for the neurons studied here, as spike rates were well below 20 Hz. The estimated input $\hat{x}$ is the first order estimate of the stimulus. The second order estimate can be obtained by convolving $\hat{x}(t)$ with h_2 and so forth for successively higher order estimates. However, for purposes of simplicity I use only the first order estimate.

The coding fraction c of a neuron is defined as 1 minus the ratio between the mean square estimation error of the stimulus normalised by the standard deviation of the stimulus amplitude:

$$c = 1 - \frac{\sqrt{< (x(t) - \hat{x}(t))^2 >_t}}{\sigma_x} \tag{5}$$

The range of the coding fractions goes from 0 (chance level predictive performance) to 1 (perfect estimation). The coding fraction increases the higher the degree of kernels used to reconstruct the neuronal stimulus, as the quality of the reconstruction is improved. Would the system be linear, the coding fraction would be, in absence of noise, unity. Note, that a coding fraction significantly greater then zero, implies that the stimulus can be (at least partially) reconstructed, thus validating the system identification model with respect to the differential equation model of the neuron or neurophysiological data itself. I assume by the design of the stimulus, that over sufficiently long stimulation periods, one will have probed a neurons response to "almost every possibly" stimulus.

3 Results

Simulations of the three neurons in response to identical stimulation showed that they had different average spike response rates (see Tab. 1). Epilepsy is typically associated with an increase in neuronal excitability and spontaneous activity and one would expect channelopathies to drive neurons to increased average spike rates. Yet, the neuron with the W1204R mutant produced actually lower response rates then the wild-type neuron. In contrast, the R1648 mutant, produced more then twice as high response rates then the wild-type, as one would have expected.

Neuron	h_0
NaV 1.1 wild-type	7 Hz
NaV 1.1 R1648H	16.8 Hz
NaV 1.1 W1204R	5.6 Hz

Table 1. Zero-th order kernel (h_0), average spike rate for the three neuron types, data averaged over $300s$.

The first-order kernel h_1 (see Fig. 3) for the three neurons differed. The wild-type and W1204R mutant (black and blue curves) produced a pronounced peak close to spike initiation ($t = -5$ to 0 ms) and in the intermediate region ($t = -20$ to -5 ms) an extensive, but flat negative region. The peak is interpretable in terms of the inputs as a type of integrator property, i.e. if inputs are strongly positive for 5 ms a spike is going to be initiated. In contrast the negative region, suggests, that positive excitation for periods between -20 and -5 ms are reducing the likelihood of a spike being initiated. Thus, these two neurons appear to be sensitive to transitions from negative to positive synaptic input drive (thereby acting like a differentiator). The negative region in the W1204R mutant is deeper, requiring the negative synaptic drive period to last slightly longer to produce a spike. This difference may account for the lower average spike rate in contrast to the wild-type.

In line with what would be expected from the h_0 kernel, the R1648 mutant neuron differed markedly (red curves) from the wild-type (and the W1204R mutant): The peak close to spike initiation is about half-as tall, i.e. half the amount of short-term positive synaptic drive is required to produce a spike. This effect possibly accounts alone for the overall higher excitability (and average spike rate) of this mutant. However, a second marked difference arises, in the region between $t = -20$ and $t = -5$ ms, where a second positive excitation peak, albeit about a tenth in amplitude as the main peak, is present. This suggests, that the R1648 mutant response characteristics are less that of differentiator-integrator like the wild-type, but more to rapid successions of positive-negative-positive synaptic input, i.e. oscillations.

The second order kernels h_2 for the three neurons (see Fig. 4.A,B,C) show similar pairwise differences (subfigure D,E). The wild-type neuron is sensitive

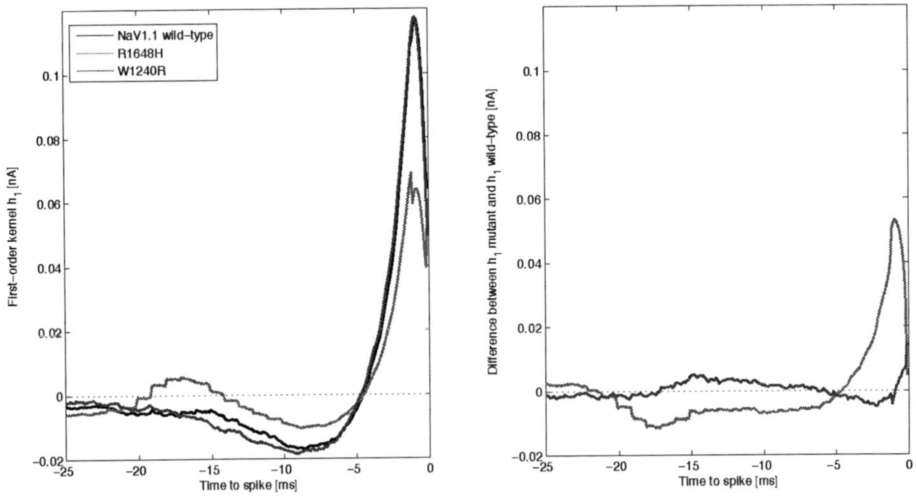

Fig. 3. (Left) First order h_1 kernels for the mammalian sodium channel NaV1.1 (Wild-type, black) and two known mutants W1204R (blue) and R1648H (red). (Right) Pair-wise differences between the mutant and the wild-type h_1 kernel.

to integrator like pairs of excitatory inputs if they occur within 5 ms and to differentiator-like pairs of inputs, if they occur within 5 ms (excitatory, yellow-red colors) and 5 to 17 ms (inhibitory, blue- colors) time-windows. While the neuron's spike response was indifferent to intermediate pairings of inputs, e.g. at 10 and 20 ms. The general picture remains the same for both mutants, however the extent to which inhibitory synaptic drive facilitates spiking differs. The R1648 mutant has a much narrower region where it is sensitive to differentiator-type inputs (5 to 20 ms for inhibition and 0 to 2 ms for excitation). In contrast, the W1204R mutant's sensitivity is similar to that of the wild-type, the differentiator region is smaller and (5 to 13 ms for inhibition and 0 to 4 ms for excitation overall weaker then the wild-type.

Neurons	Ratio
NaV 1.1 W1204R over NaV 1.1 wild-type	93%
NaV 1.1 R1648H over NaV 1.1 wild-type	78%

Table 2. Ratios of the (first-order) coding fraction for the three neurons (see methods for details).

Thus, point mutations in sodium channels produce detectable differences in overall input-output relationships of three neurons that are evident in each of the kernels calculated here. The coding fraction – based on the first-order, linear estimation of the stimulus from the spike train – provides further insight into the

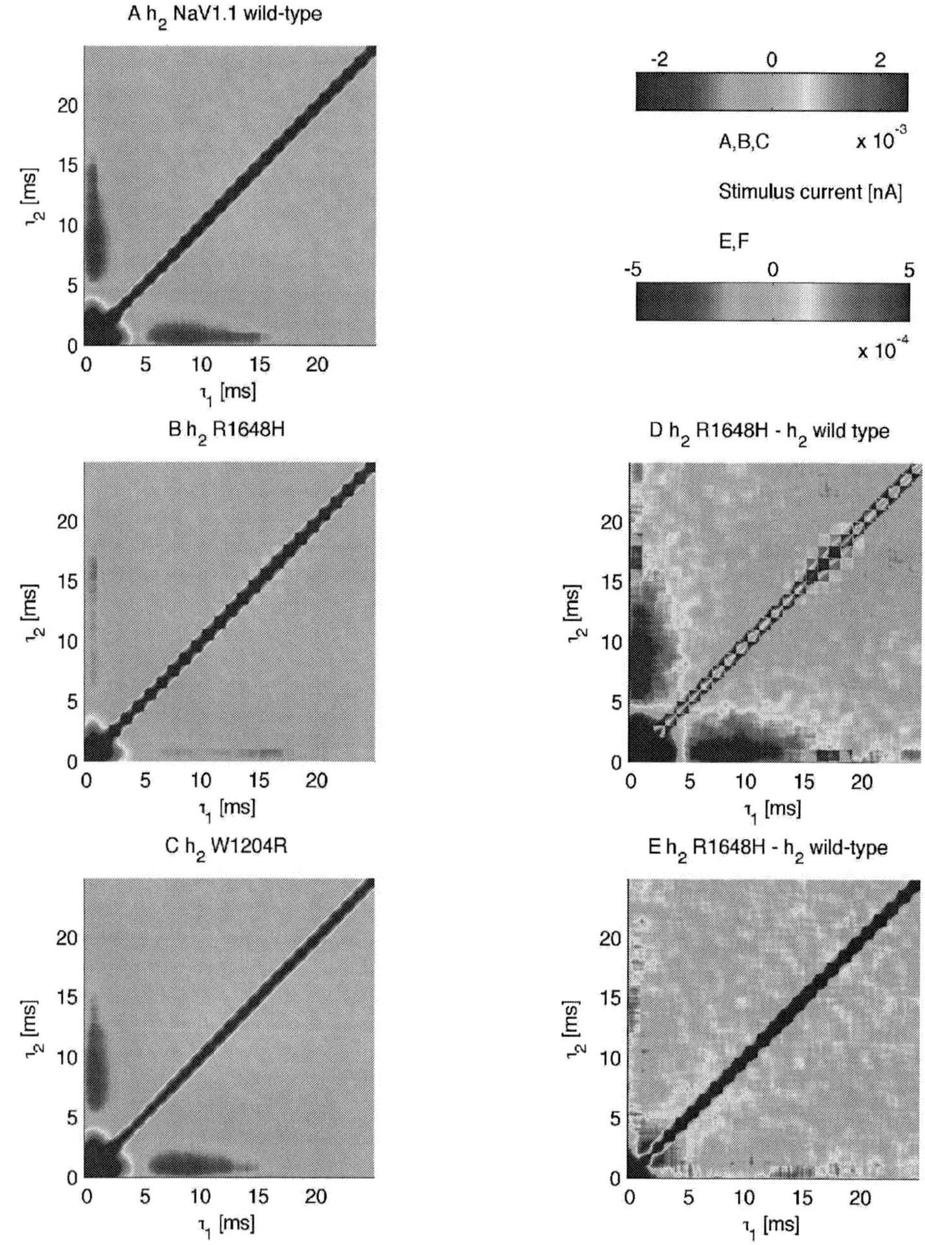

Fig. 4. Second-order $h_2(\tau_1, \tau_2)$ kernels for neurons with (A) NaV1.1 (Wild-type, black) and two mutant channels, (B) W1204R and (C) R1648H. Pairwise differences between h_2 of (D) wild-type and W1204R and (E) wild-type and R1648H. Color bars top right indicate the scale for the corresponding stimulus current colors, top bar for subfigures A-C and bottom bar for subfigures D,E.

differences caused by the mutation. The ratios of the mutant neuron's coding fraction over that of the wild-type neuron provide a measure by how much the (non-linear) response properties changed (see Tab. 2). While the W1204R mutant was about as non-linear as the wild-type, the R1648H displayed a considerable deviation, suggesting that it is reacting to more complex input patterns.

4 Discussion

I showed here how measures of neuronal performance derived from well-known system identification techniques can be used to assess the functional impact of channelopathies and their effects on the network-level. The first- and second-order Wiener kernels and the coding fraction provide an intuitively interpretable measure as to how channelopathies alter neuronal behaviour and performance.

Many neurons in mammalian central nervous system, receive inputs from 1000s of synapses. This large and dynamic synaptic patterns is not random but highly structured[24]. The system identification methods used here demonstrate how reaction to identical synaptic input patterns can be very different in mutant neurons and that mutant neurons may start to react to different synaptic inputs from the wild-type (cf h_1 of wild-type and R1648H).

The kernel methods used here can provide insight to what extent low-level changes, for example in inactivation kinetics will impact system level response to inputs. Normally wild-type Na^+channels open in response to membrane depolarization from excitatory inputs, then enter inactivation states that prevent reopening during sustained depolarization [25]. Many Na+ channel mutations linked to clinical syndromes exert are known to affect the inactivation process either by reducing the rate of inactivation or increasing the rate of recovery from inactivation, as is the case for the W1204R mutant.This is generally believed to cause an increase in excitability and thus facilitation of seizures [26]. Yet, overall neuronal behaviour, as calculated from the Wiener kernels calculated here and compared to wild-type and other mutants, suggest that this assumed effect of inactivation can be much smaller (cf W1204R and wild-type coding fractions) when considering naturalistic synaptic inputs and network-level behaviour. However, it could be that we have not probed specific patterns of synaptic inputs, that may promote the impact of W1204R - using the Wiener kernels as templates, would enable us to search for such specific input patterns.

The fact that many neurons and axons in the nervous system are very small, implies that channel noise effects have to be taken into account when studying channelopathies [10,11]. Especially, because channel noise can by itself, produce random spontaneous spikes, which may be important with respect to seizure-facilitating channel mutations [10]. This feasibility study here ignored noise, both present in ion channels as well as other parts of the neuron (Faisal, Selen & Wolpert, in Review, Nature Reviews Neuroscience). However, this approach outlined here could easily be extended to account for such stochastic effects, e.g. by running the simulation in stochastic ion channel gating mode (see [11]) and re-applying the same analysis methods. This would require longer simulations, of about 20-30 minutes of simulated time. Coupled with large-scale computer simulations the inclusion of noise should yield sufficient data (>2 hour of simulated time) for the application of information theory [16], as another complimentary measure of neuronal performance.

Previous important physiological studies investigating the link between channelopathies and neuronal behaviour used methods that allowed little insight into network-level impact, due to three key weaknesses (e.g. [17,18,7,8]). First, the use of non-physiological conditions, for example voltage-clamping the neuron to impose an action potential wave form instead of using a neuron the can generate action potentials by itself. I addressed this here by embedding the mutant ion channels in a model neuron which uses membrane properties and delayed rectifier K^+channels that allow a wide variety of Na channels (NaV1.1,NaV1.2,NaV1.6 and mutants thereof; data not shown) to generate spiking responses. Thus, allowing for a common framework to compare the effects of mutations on a large variety of sodium and potassium channel mutations. Second, unnatural stimulation protocols are often used, such as constant, step or ramp-like input stimuli that have little to do with synaptic stimuli encountered *in vivo*. I addressed this by using fluctuating current stimuli (which are also required for the system identification technique used here) that mimic naturalistic synaptic drive [27]. Thirdly, and most importantly, measures of neuronal performance were previously used which are known to be at best only weakly capturing the nature of the neural code [16] or will be only interpretable for highly specific cases, such as counting the average spike rate in response to constant current stimuli or counting the number of spikes in ramp stimuli. Thus, system-level impact of channelopathies was previously not quantified.

The system identification approach presented here addresses the macroscopic response properties of neurons, and translates the impact of the specific low-level cause of pathological behaviour to the whole-cell and network level. Given the non-linearity of the action potential mechanism, the response kernel approach may provide valuable insight into what cell parameters could be adjusted by pharmacological means to compensate for changes in response properties caused by ion channel mutations. Previous low-level methods suggest that one should try to reverse the channelopathies' impact at the molecular level, e.g. by focusing on the mutated channel itself, because it will be unclear how altering other parameters will affect whole-cell behaviour. In contrast using the system identification approach offers more degrees of freedom to overcome the system-level effects of channelopathies. I can now assess how varying every possible adjustable parameter (e.g. altering potassium channel density through channel blockers, shifting membrane resting potential through dietary regimes) will restore "healthy" whole-cell behaviour, instead of solving the much harder task to restore "healthy" ion channel behaviour. This system-level focus is important because homeostatic plasticity mechanisms [28,29] can partially compensate for deviations in individual cell properties. It may be thus possible to make advantageous use of homeostatic plasticity mechanisms for clinical treatment, but only if one understands how whole-cell behaviour is affected. Thus, for the brain to function as a whole, not the molecular behaviour, but the resulting whole-cell behaviour is key.

While, I showed here that the feasibility of system identification techniques on voltage-gated Na^+channels, the analysis described above is also straightforward applicable to voltage-gated and Ca^{++}-gated K^+channels. Such channels have been found to be linked to genetic mutations in neuronal potassium channels, in particular those in the KCNQ family [1]. Further work on quantitative approaches of mutation-altered ion channel gating is bound to improve the understanding of molecular mechanisms of disease phenotypes. This approach will eventually allow us to compare and link the genetic sequence of the ion channel and in the physiological domain, it's kinetics, with the overall response behaviour of the neurons.

Acknowledgments

AAF was supported by the Wellcome Trust and a Research Fellowship by Wolfson College in the University of Cambridge.

References

1. Jentsch, T.: Neuronal kcnq potassium channels: physiology and role in disease. Nat. Rev. Neurosci. **1**(1) (2000) 21–30
2. : (Resurgence of sodium channel research)
3. Isom, L.: Beta subunits: players in neuronal hyperexcitability? Novartis Found Symp. **241:124-38** (2002)
4. Wallace, R., Scheffer, I., Parasivam, G., Barnett, S., Wallace, G., Sutherland, G., Berkovic, S., Mulley, J.: Generalized epilepsy with febrile seizures plus: mutation of the sodium channel subunit scn1b. Neurology **58**(14) (2002) 1426–1429
5. Noebels, J.: Sodium channel gene expression and epilepsy. Novartis Found Symp. **241:109-20** (2002)
6. Castaldo, P., del Giudice, E., Coppola, G., Pascotto, A., Annunziato, L., Taglialatela, M. (J.Neurosci.)
7. Spampanato, J., Escayg, A., Meisler, M., Goldin, A.: Generalized epilepsy with febrile seizures plus type 2 mutation W1204R alters voltage-dependent gating of Na (v) 1.1 sodium channels. Neuroscience **116**(1) (2003) 37–48
8. Spampanato, J., Aradi, I., Soltesz, I., Goldin, A.: Increased Neuronal Firing in Computer Simulations of Sodium Channel Mutations That Cause Generalized Epilepsy With Febrile Seizures Plus. Journal of Neurophysiology **91**(5) (2004) 2040–2050
9. Laughlin, S.B., Sejnowski, T.J.: Communication in neuronal networks. Science **301**(5641) (2003) 1870–1874
10. Faisal, A.A., White, J.A., Laughlin, S.B.: Ion-channel noise places limits on the miniaturization of the brain's wiring. Curr Biol **15**(12) (2005) 1143–1149
11. Faisal, A., Laughlin, S.: Stochastic simulations on the reliability of action potential propagation in thin axons. PLoS Computational Biology **3**(5) (2007) e79 doi:10.1371/journal.pcbi.0030079

12. Hodgkin, A., Huxley, A.: Quantitative description of membrane current and its application to conduction and excitation in nerve. J. Physiol. (London) **117** (1952) 500–544
13. Wiener, N.: Cybernetics or control and communication in the animal and the machine. 2nd (ninth printing of 1999) edn. MIT Press, Cambridge, MA (1961)
14. Marmarelis, P., Marmarelis, V.: Analysis of PHysiological systems – The White-Noise Approach. Computers in Biology and Medicine. Plenum Press, New York (1978)
15. Bryant, H., Segundo, J.: Spike initiation by transmembrane current: a white-noise analysis. J. Physiol. (London) **260** (1976) 279–314
16. Rieke, F., Warland, D., de Ruyter van Steveninck, R.R., Bialek, W.: Spikes : Exploring the neural code. Computational neuroscience. MIT Press, Cambridge (MA) (1997)
17. Escayg, A., MacDonald, B., Meisler, M., Baulac, S., Huberfeld, G., An-Gourfinkel, I., Brice, A., LeGuern, E., Moulard, B., Chaigne, D., et al.: Mutations of SCN1A, encoding a neuronal sodium channel, in two families with GEFS+ 2. Nature Genetics **24** (2000) 343–345
18. Escayg, A., Heils, A., MacDonald, B., Haug, K., Sander, T., Meisler, M.: A novel SCN1A mutation associated with generalized epilepsy with febrile seizures plus and prevalence of variants in patients with epilepsy. Am J Hum Genet **68**(4) (2001) 866–873
19. Rudolph, M., Destexhe, A.: Characterization of subthreshold voltage fluctuations in neuronal membranes. Neural Comput **15**(11) (2003) 2577–618
20. Koch, C.: Biophysics of computation. Computational neuroscience. Oxford University Press, Oxford (1999)
21. Faisal, A., Laughlin, S., White, J.: How reliable is the connectivity in cortical neural networks? In Wunsch, D., ed.: Proceedings of the IEEE Intl. Joint. Conf. Neural Networks 2002, INNS (2002) 1–6
22. Bialek, W., Rieke, F., de Ruyter van Steveninck, R., Warland, D.: Reading a neural code. Science **252** (1991) 1854–1857
23. Gabbiani, F., Koch, C.: Coding of time-varying signals in spike trains of integrate-and-fire neurons with random threshold. Neural Computation **8**(1) (1996) 44–66
24. Softky, W.R., Koch, C.: The highly irregular firing of cortical cells is inconsistent with temporal integration of random epsps. J Neurosci **13**(1) (1993) 334–50
25. Hille, B.: Ion channels of excitable membranes. 3rd edn. Sinauer Associates, Sunderland, MA (2001) 814 pp.
26. Catterall, W.: From Ionic Currents to Molecular Mechanisms The Structure and Function of Voltage-Gated Sodium Channels. Neuron **26**(1) (2000) 13–25
27. Schneidman, E., Freedman, B., Segev, I.: Ion channel stochasticity may be critical in determining the reliability and precision of spike timing. Neural Computation **10** (1998) 1679–1703
28. Marder, E., Prinz, A.A.: Modeling stability in neuron and network function: the role of activity in homeostasis. Bioessays **24**(12) (2002) 1145–54
29. Niven, J.E., Vahasoyrinki, M., Kauranen, M., Hardie, R.C., Juusola, M., Weckstrom, M.: The contribution of Shaker K+ channels to the information capacity of Drosophila photoreceptors. Nature **421**(6923) (2003) 630–634

Using simple molecular orbital calculations to predict disease: fast DFT methods applied to enzymes implicated in PKU, Parkinson's disease and Obsessive Compulsive Disorder.

Laura Hofto[1], Meghan Hofto[1,2], Jessica Cross[1], and Mauricio Cafiero[1]

[1]Department of Chemistry, Rhodes College, 2000 North Parkway, Memphis, TN 38112
[2]College of Medicine, University of Tennessee Health Science Center, 930 Madison Ave, Memphis, TN 38163

Abstract. Many diseases can be traced to point mutations in the DNA coding for specific enzymes. These point mutations result in the change of one amino acid residue in the enzyme. We have developed a model using simple molecular orbital calculations which can be used to quantitatively determine the change in interaction between the enzyme's active site and necessary ligands upon mutation. We have applied this model to three hydroxylase proteins: phenylalanine hydroxylase, tyrosine hydroxylase, and tryptophan hydroxylase, and we have obtained excellent correlation between our results and observed disease symptoms. Furthermore, we are able to use this agreement as a baseline to screen other mutations which may also cause onset of disease symptoms. Our focus is on systems where the binding is due largely to dispersion, which is much more difficult to model inexpensively than pure electrostatic interactions. Our calculations are run in parallel on a sixteen processor cluster of 64-bit Athlon processors.

1 Introduction

1.1 Protein structure and genetics

Modeling enzymes and enzyme active sites for the purpose of identifying molecular-level causes of disease is a difficult and complicated business. The first difficulty involves the structure of the enzyme itself[1]. DNA and amino acid sequence information about enzymes is fairly easily attainable today, but, since structure does not necessarily follow from sequence in a predictable manner, more information is necessary. Protein homology is a commonly used tool to obtain a good guess as to the general structural motifs in a novel protein, but it provides no real quantitative structural information. The two best tools for ascertaining qualitatively meaningful structures for enzymes are nuclear mag-

CP940, *CompLife 2007, 3rd International Symposium on Computational Life Science*
edited by A. Siebes, M. Berthold, R. Glen, and A. Feelders

netic resonance (NMR) techniques and crystallography techniques. These techniques provide a set of *xyz* coordinates for all heavy atoms in a protein and for associated solvent molecules within some resolution on the angstrom level. Since our work depends on knowing an enzyme's structure, we focus on proteins with a known crystal structure or reliable NMR structure. We assume this structure is qualitatively correct within the experimental resolution; in particular, the amino acid backbone is assumed to be rigid in most of our work. Furthermore, in most cases we require a crystal structure including bound ligands associated with the condition we are studying.

Knowing the enzyme structure, we now focus on diseases associated with point mutations causing a change in one amino acid residue in this enzyme. Specifically, we are interested in point mutations affecting residues in the ligand binding-active site. There is much work in the area of intermolecular forces binding ligands to active sites, and in this work we focus specifically on aromatic ligands which are bound by aromatic residues. While covalently bound ligands and electrostatically (and hydrogen-bond) bound ligands are more straightforward to investigate using conventional techniques, the dispersion-dominated systems we examine require a more careful approach.

1.2 Computational Chemistry

Commonly used methods for investigating protein/ligand interactions where electrostatics are the most important forces include classical molecular mechanics methods, such as those employed in popular modeling packages such as AMBER[2,3], or, occasionally, semi-empirical quantum mechanical methods such as AM1[4], PM3[5], and others. These classical and semi-empirical methods do an excellent job in modeling electrostatically dominated systems, but when applied to dispersion problems, these methods are usually supplemented by an empirical "dispersion potential" which is often calibrated to work with the system of interest.

Rigorous, quantitative treatment of dispersion requires use of a quantum mechanical method which takes into account electron correlation [6]. The basic molecular orbital-based quantum mechanical method used in chemistry applications is the Hartree-Fock (HF) method. This method assumes several things. First, the electrons are assumed to move in a field of static nuclei (the Born-Oppenheimer, or clamped-nucleus approach). HF also assumes that the molecular wave function (ψ) can be expressed as a single determinant of molecular orbitals (ϕ):

$$\psi(r_1, r_2, \ldots r_n) = \tfrac{1}{n!}\det[\phi_1(r_1)\phi_2(r_2)\ldots\phi_n(r_n)], \quad (1)$$

$$\phi_i(r_i) = \sum c_j \chi_j(r_i) \quad (2)$$

In practice, the molecular orbitals are usually expressed in a basis set of auxiliary functions (χ), commonly Gaussian functions, but sometimes exponential, or Slater-type functions. The HF model also assumes that each electron 'sees' the forces due to the other electrons only in an averaged sense. This approximation causes HF to lose electron correlation, or the instantaneous interactions of electrons, which is necessary to model dispersion. Corrections must be made to the HF model in order to add correlation back in, and these methods are referred to as post-HF, or correlated methods. Of these methods, the most inexpensive method which provides meaningful data is the second order Moller Plessett theory (MP2). This method considers contributions to the molecular energy and wave function due to doubly excited determinants:

$$\psi_{MP2}(r_1, r_2, \ldots r_n) = \psi_{HF} + \sum_{i<j} \sum_{a<b} C_{ij}^{ab} \Phi_{ij}^{ab} \tag{3}$$

There are more accurate and expensive post-HF methods available, such as the near-industry standard coupled cluster singles, doubles, and perturbative triples [CCSD(T)], but these are infeasible for application to large protein/ligand complexes. Thus in our work, we use the MP2 method as a standard, and consider those answers correct. We must keep in mind, though, that it not as accurate as CCSD(T), but in this work we consider *differences* in energies, rather than absolute energies, and so the inaccuracy is not as serious an issue as it could be.

Another major focus of our work is in identifying inexpensive, single-determinant methods. By single determinant methods, we mean methods which, like HF, do not make explicit reference to excited states and thus cost about as much time to run as HF, compared with the increased cost of MP2, which can take orders of magnitude more time. These methods are called Density Functional theory (DFT) methods [7]. The DFT methods (specifically the Kohn-Sham implementation of DFT which we use [8]) use an exchange/correlation potential, v_{XC}, to add electron correlation energy within a single determinant framework[9];

$$E[\rho, \nabla\rho] = E_T[\rho] + E_J[\rho] + E_{ext}[\rho] + E_{XC}[\rho, \nabla\rho] \tag{4}$$

$$\rho = 2\sum \phi_{KS}^2 \tag{5}$$

here, ρ is the electron density obtained from summing over the squares of the Kohn-Sham orbitals, ϕ_{KS}, and the subscripts in Eq. 4 refer to T, the kinetic energy, J, the coulomb energy, ext, the external potential, and XC, the exchange/correlation energy. While Kohn-Sham DFT is a formally exact theory, the form of the XC potential, and thus energy, is not known, and so different approximations are used for correlation. These correlation functionals (functions of the electron density) are often found to work well in some cases and not others. We hope to find DFT methods which perform well in the protein/ligand complexes we study and then use them to design novel methods which are more generally transferable. These DFT methods would be an excellent compromise in the study of large biochemical systems as they are relatively inexpensive and relatively accurate.

We use then three types of methods to study the interactions between active site residues and ligands: HF (which does a good job of predicting electrostatic interactions), MP2 (which adds dispersion in a rigorous way), and DFT (which adds dispersion in a non-predictable way). We study the MP2 results to provide answers to our questions about molecular causes of disease, and we study the DFT methods in hopes of finding a method of arriving at these answers in a less expensive way in the future. Of the three DFT methods we use, one, SVWN, is a *local* method, which uses only the density, ρ, in the XC potential and energy; a second, HCTH407, is a *semi-local* method, which uses the density and the density gradient in the XC term; the last DFT method, B3LYP, uses the density, the gradient of the density, and the non-local HF exchange potential/energy.

1.3 Biochemistry

The hydroxylase enzymes we study are all involved in the production of neurologically important molecules and they are all somewhat homologous. Phenylalanine hydroxylase (PheOH), tyrosine hydroxylase (TyrOH), and tryptophan hydroxylase (TrpOH), all have similar active sites that bind several ligands, including tetrahydrobiopterin (BH4). BH4 is used in all the enzymes to hydroxylate the substrate. Figure 1 shows that alll of these enzymes feature a phenylalanine residue which forms a Pi-stacking complex, or sandwich complex, with the BH4 (and other aromatic ligands). Mutations of this residue in PheOH (F254I) are known to cause the disease Phenylketanuria (PKU)[10]; mutations in the analogous residue in TyrOH (F300A) are known to cause a decrease in the production of L-DOPA (a condition implicated in Parkinson's disease)[11,12]. Mutation in the analogous residue in TrpOH should also cause problems in the production of serotonin, a condition implicated in anxiety and obsessive compulsive disorder.

We compute interaction energies between active site residues and ligands for wildtype enzymes, and then we perform *in silico* mutations of the residues to see how the interactions change upon mutation. By using the interaction energies of the wildtype structures and known disease-inducing mutations, we can calibrate our model chemistry. It is then possible to perform additional mutations and make predictions as to whether or not they will also cause disease symptoms. Our model assumes that the mutations studied do not change the general shape of the active site considerably.

The majority of our calculations are performed using the Parallel Quantum Solutions (PQS) software [13], and run on a PQS-built sixteen processor Beowulf cluster running Linux.

2. Protein/Ligand Model Chemistry

We obtain crystal structures for the enzymes used in our study from the Brookhaven protein databank [10,11,12]. We isolate the aromatic protein residues of interest bound in complexes with ligands using Arguslab[14]. The complexes studies include: Phe254/BH4 and Tyr325/BH4 (both from PheOH), Phe300 and Phe309 (both from TyrOH), and Phe241 (from TrpOH). The amino acid residues are 'capped' to create neutral amino acids by adding oxygen atoms or hydroxyl groups where appropriate. We form neutral amino acids in order to minimize electrostatic interactions and emphasize the desired dispersion reactions. Protons are also added to the complexes as appropriate to maintain neutrality. We use a modest level of computational theory to optimize the positions of the added protons and oxygen atoms/hydroxyl groups, B3LYP [15]/6-31G. We choose this level because B3LYP is an inexpensive DFT method which does an excellent job of finding equilibrium structures of organic molecules. Adding additional polarization or diffuse functions to the basis set for optimization is not necessary, as the weakly-orbital dependant B3LYP method does not improve structures considerably with this addition. We do not use solvent models in this optimization, as the presence of the solvent is implicit in the NMR or crystal structures from which we begin our investigations.

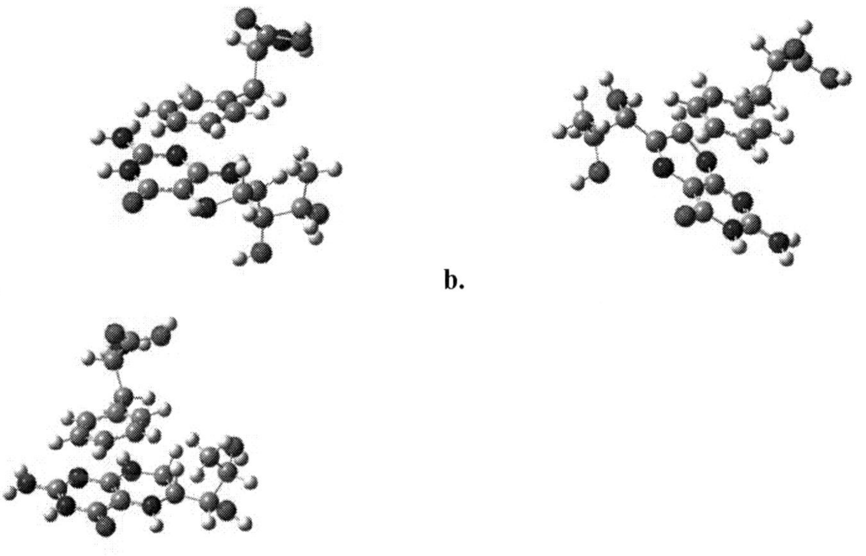

a. b.

c.

Fig. 1. The aromatic Phe residue in the three enzymes, phe254 in PheOH (a), phe300 in TyrOH (b), and phe241 in TrpOH (c), shown in a Pi-stacking configuration with the BH4 molecule.

We find the interaction energy between the residue and the ligand using three DFT methods: B3LYP, SVWN [16,17], and HCTH407[18,19] using the 6-311++G** basis set[20], and HF and MP2 using the 6-311G** basis set[21]. We compute the energies of the complex and the monomers, and apply the counterpoise correction to account for basis set superposition errors [22].

We perform *in silico* mutations on the wildtype residues by removing the side chains and replacing them with new side chains. The positions of the new atoms are optimized again using the B3LYP/6-31G level of theory. The interaction energies between the mutant ligands and BH4 are computed as above.

3. Hydroxylase enzymes and molecular causes of disease: MP2 results

Our recent work has focused on aromatic residue/ligand interaction in tryptophan hydroxylase. TrpOH hydroxylizes tryptophan into 5-hydroxy-tryptophan (5HT) using BH4 as a cofactor; 5-hydroxy-tryptophan is then made into serotonin by a decarboxylase enzyme. In this work, we focus on TrpOH's ability to bind BH4 for use in the hydroxylation reaction. A mutation in TrpOH which could cause reduced interaction with BH4 would slow the production of serotonin. Lack of sufficient serotonin is implicated in such conditions as depression, obsessive compulsive disorder, migraine, bipolar disorder, and anxiety.

The TrpOH residue Phe241 forms a sandwich-type pi-complex with BH4 (see Figure 1); this complex has a counterpoise-corrected MP2 interaction energy of about 3.6 kcal/mol. While this is not a large interaction, we can use our knowledge of disease causing mutations in the related enzymes PheOH and TyrOH to establish a correlation between interaction energies and loss of enzyme function. We know that changes to the analogous residues which aid in the aromatic binding of BH4 to PheOH and TyrOH can cause a serious loss of enzymatic function. These residues are Phe254 in PheOH and Phe300 in TyrOH. The interaction energies between BH4 and the aromatic Phe residues in those enzymes are about 5-6 kcals/mol, so the forces in play are comparable; if point mutations to this one rsidue can cause disease in PheOH and TyrOH, then we can expect similar results in TrpOH. Table 1 shows the counterpoise-corrected interaction energies between BH4 and wildtype and various mutant residues at the 241 position.

In order to make predictions concerning TrpOH, we calibrate our methods using known phenotype/computation correlations from our previous work on PheOH and TyrOH[23]. Tables 2 and 3 show the counterpoise-corrected interaction energies between BH4 and the wildtype and mutant residues for both PheOH and TyrOH. In TyrOH, we know that the wildtype (Phe300) has full function in the production of L-DOPA, and it has been shown that a mutation to alanine at that site (F300A) causes an ~40% drop in L-DOPA production. We can see from our data that if we take the wildtype as having 100% interaction, the mutation to alanine causes a drop to 23% interaction. We thus can establish two points on a correlation curve between interaction and enzyme function. In PheOH, we know that the wildtype residue (Phe254) has full enzyme function, and a mutation to isoleucine (F254I) causes symptoms of PKU. Again our data shows that this mutation causes a drop in interaction from 100% to 22.9%. Thus we add another two points to our correlation and we see that 23% is a common point at which enzyme function will fail.

Table 1. Counterpoise corrected interaction energies between BH4 and the residues at position 241 in TrpOH in kcals/mol. DFT calculations use the 6-311++G** basis set, and HF and MP2 calculations use the 6-311G** basis set.

BH4 with	HF	B3LYP	HCTH407	SVWN	MP2	%
Phe241	8.23	6.01	5.26	-4.76	-3.60	100
Ala241	0.79	0.58	-0.70	-3.00	-2.09	58
Leu241	7.21	4.88	5.80	-7.42	-3.22	89
Ile241	12.76	8.56	12.04	-8.06	-0.16	4
Gly241	-0.23	-0.05	-0.95	-0.57	-0.88	24
Tyr241	8.50	6.18	3.35	-4.63	-3.48	97
Gln251	1.30	1.60	2.10	-7.10	-4.00	109

Table 2. Counterpoise corrected interaction energies between BH4 and the residues at position 300 in TyrOH in kcals/mol. DFT calculations use the 6-311++G** basis set, and HF and MP2 calculations use the 6-311G** basis set.

BH4 with	HF	B3LYP	HCTH407	SVWN	MP2	%
Phe300	5.18	3.64	2.65	-5.43	-6.43	100
Ala300	0.82	0.49	-0.58	-1.92	-1.49	23
Leu300	12.70	8.93	10.8	-4.19	-0.10	1
Ile300	4.39	2.76	3.00	-6.39	-4.29	67
Gly300	-0.30	-0.28	-0.77	-0.57	-0.84	13
Met300	4.97	3.1	3.63	-7.03	-4.46	70
Tyr300	5.50	3.51	2.35	-5.71	-6.87	107

Table 3. Counterpoise corrected interaction energies between BH4 and the residues at position 254 in PheOH in kcals/mol. DFT calculations use the 6-311++G** basis set, and HF and MP2 calculations use the 6-311G** basis set.

BH4 with	HF	B3LYP	HCTH407	SVWN	MP2	%
Phe254	5.61	3.76	2.84	-6.44	-5.54	100
Ala254	0.71	0.58	-0.90	-2.27	-1.53	27
Leu254	3.14	2.20	0.97	-5.12	-3.55	64
Ile254	6.71	4.83	4.97	-4.95	-1.27	22
Gly254	0.00	0.10	-0.65	-0.33	-0.59	10
Met254	0.73	0.32	-1.47	-5.26	-4.27	77
Tyr254	6.12	4.34	3.15	-5.59	-5.02	90

Using this information from our previous work, we are in position to make predictions in our current work on TrpOH. Table 1 shows that, of the mutations studies, the two cases similar to the 'calibration points' in PheOH and TyrOH (mutations to alanine and isoleucine) both produce the expected drop in interaction energy (58% for alanine and 4% for isoleucine). The alanine mutation corresponds very well with the previous known drop in production of L-DOPA in TyrOH, although it is not as dramatic. The reduction in interaction for the isoleucine mutation, though, agrees perfectly with the data for PKU caused by PheOH, and we can say that this mutation will likely cause a reduction in the production of 5HT, and in turn, a reduction in serotonin. This could lead to onset of symptoms mentioned above, such as anxiety, depression, and OCD. Of the other mutations studied, the leucine, tyrosine, and glutamine mutants would likely retain the interaction energy necessary to keep protein function, and the glycine mutant would lose 76% of the interaction energy.

4. Analysis of DFT methods for dispersion bound ligands

All of the above discussion is based on MP2, which provides information about electrostatic interaction plus a rigorous approximation of dispersion. We also studied HF, which does not include dispersion, and three DFT methods, which include dispersion in an approximate way. We can see by the data in Table 1 that HF does not find attractive interactions in any case except glycine, and in fact finds large repulsive interactions. Glycine is a small residue and interacts almost exclusively by electrostatics, and this is why HF performs decently. Of the DFT methods studied, B3LYP is the most popular, and is widely used by non-specialists. We can see that B3LYP is both quantitatively and qualitatively inaccurate. It finds only repulsive interactions of an almost random nature (except glycine). This result, the failure of B3LYP to accurately model systems where dispersion-like forces are a major contributor, is consistent with much of our previous work on aromatic-type systems[23,24,25,26,27]. HCTH407, which has worked well in some of our recent applications in dispersion bound biomolecules, also finds largely repulsive interactions.

As in our past work on PheOH and TyrOh, we find that SVWN is the only DFT method studied that performs well for these systems, following the MP2 trends fairly well, although it almost always overestimates the interaction energies. Since SVWN is very inexpensive compared with MP2 and in some cases can be coded to scale linearly with the size of the system, it is a good alternative to the MP2 calculations for qualitative predictions of changes in interaction energy upon mutations.

5. Conclusions

We show that our simple molecular orbital model describing the effects of point mutations in enzyme active sites is an effective way of screening possible mutations for deleterious effects on protein function. We also show that the inexpensive SVWN method closely approximates MP2 in many cases and may be used as an effective screening tool. We have largely ignored solvation in our calculations. We believe that solvation would change our results only minimally since the crystal structures we begin with are derived from once-solvated structures. Further, unpublished work in our lab had explored the effect of including larger basis sets and implicit solvation methods on the geometry optimization methods we use, and the differences with the methods we do use are minimal.

References

1. For an excellent overview of protein structures in molecular modelling, see: Rappe, A.K., and Casewit, C.J., *Molecular Mechanics across Chemistry*, University Science Books, 1997.
2. Case, D.A., Cheatham, T.E., III, Darden, T., Gohlke, H., Luo, R., Merz, Jr., K. M., Onufriev, A., Simmerling, C., Wang, B., and Woods. R., *J. Computat. Chem.* **26,** 1668-1688 (2005).
3. Ponder J. W., and Case, D. A., *Adv. Prot. Chem.* **66,** 27-85 (2003).
4. Dewar, M. J. S., Zoebisch, E. G., Healy, E. F. and Stewart, J. J. P., *Journal of the American Chemical Society*, **107**, 3902, (1985).
5. Stewart, J. J. P. *J. Comput. Chem.* **1989,** *10*, 209.
6. Discussion of the quantum mechanics methods described here, including the Hartree-Fock, MP2, and CCSD(T) methods may be found in any introductory computational chemistry text, such as: Jensen, F., *Introduction to Computational Chemistry*, John Wiley and Sons, 1999.
7. Hohenberg, P., and Kohn, W., *Phys. Rev.*, **136**, B864, (1964).
8. Kohn, W., Sham, L.J., *Phys Rev.*, **140**, A1133, (1965).
9. LDA, GGA, and hybrid DFT methods are discussed at length in: Koch, W., and Holthausen, M.C., *A Chemist's Guide to Density Functional Theory*, Wiley-VCH, 1999.
10. Erlandsen, H., Flatmark, T., Stevens, R. C., and Hough, E. *Biochemistry.* **37**, 15638-15646 (1998).
11. Goodwill, K.E., Sabatier, C., and Stevens, R.C., *Biochemistry*, **37**, 13437 - 13445 (1998).
12. Ellis, H.R., Daubner, S.C., McCulloch, R.I., and Fitzpatrick, P.F., *Biochemistry*, **38**, 10909 - 10914 (1999).

13. PQS version 3.1, Parallel Quantum Solutions, 2013 Green Acres Road, Fayetteville, Arkansas 72703.
14. ArgusLab 4.0.1, Mark A. Thompson, Planaria Software LLC, Seattle, WA, http://www.arguslab.com.
15. Becke, A. D. *J. Chem. Phys.* **98**, 5648-5652 (1993).
16. Slater, J.C., *Quantum Theory of Molecular and Solids. Vol. 4: The Self-Consistent Field for Molecular and Solids* (McGraw-Hill, New York, 1974).
17. Vosko, S.H., Wilk, L., and Nusair, M., *Can. J. Phys* . **58**, 1200 (1980).
18. Boese, A. D., and Handy, N. C. *J. Chem. Phys.* **114**, 5497-5503 (2000).
19. Boese, A. D., Doltsinis, N. L., Handy, N. C., and Sprik, M. *J. Chem. Phys.* **112**, 1670-1678 (1999).
20. Krishnan, R., Binkley, J. S., Seeger, R., and Pople, J.A., *J. Chem. Phys.* 72, 650 (1980).
21. Clark, T., Chandrasekhar, J., Schleyer, P. V. R., *J. Comp. Chem.* 4, 294 (1983).
22. Boys S.F., and Bernardi, F., *Mol. Phys.* **19**, 553 (1970).
23. Hofto, M.E., Cross, J.N., and Cafiero, M., *J. Phys. Chem. B,* DOI: 10.1021/jp072518w .
24. Hofto, L.R., Van Sickle, K., and Cafiero, M., *Int. J. Quantum Chem.*, DOI: 10.1002/qua.21426.
25. Hofto, M.E., Godfrey_Kittle, A., and Cafiero M., *THEOCHEM,* doi: 10.1016/j.theochem.2007.01.023.
26. Van Sickle, K., Culberson, L. M., Holzmacher, J. L., and Cafiero, M., *Int. J. Quantum Chem.* 107, 1523-1531 (2007).
27. Godfrey-Kittle, A., and Cafiero, M., *Int. J. Quantum Chem.,* 106, 2035 (2006).

DATA ANALYSIS

Efficient Analysis of Mass Spectrometry Data Using the Isotope Wavelet

Rene Hussong[1], Andreas Tholey[2] and Andreas Hildebrandt[1]
[1] Center for Bioinformatics,
Computer Science Department, Saarland University,
66041 Saarbrücken, Germany.
{rene,anhi}@bioinf.uni-sb.de
[2] Center for Bioinformatics,
Institute of Biochemical Engineering,
Functional Proteomics Group, Saarland University,
66041 Saarbrücken, Germany.
a.tholey@mx.uni-saarland.de

Abstract. Mass spectrometry (MS) has become today's de-facto standard for high-throughput analysis in proteomics research. Its applications range from toxicity analysis to MS-based diagnostics. Often, the time spent on the MS experiment itself is significantly less than the time necessary to interpret the measured signals, since the amount of data can easily exceed several gigabytes. In addition, automated analysis is hampered by baseline artifacts, chemical as well as electrical noise, and an irregular spacing of data points. Thus, filtering techniques originating from signal and image analysis are commonly employed to address these problems. Unfortunately, smoothing, base-line reduction, and in particular a resampling of data points can affect important characteristics of the experimental signal. To overcome these problems, we propose a new family of wavelet functions based on the isotope wavelet, which is hand-tailored for the analysis of mass spectrometry data. The resulting technique is theoretically well-founded and compares very well with standard peak picking tools, since it is highly robust against noise spoiling the data, but at the same time sufficiently sensitive to detect even low-abundant peptides.

1 Introduction

Today, mass spectrometry (MS) is one of the fundamental techniques in proteomics research. Coupled to high performance liquid chromatography (HPLC), it even allows for the comparison of whole proteomes of different cell lines. In a typical experimental setting, the probe of interest is digested with a protease (e.g. trypsin) leading to a set of smaller peptides which enter a chromatographic column. Peptides elute from the column at different time points, according to their different physical and chemical properties. After this separation step, the probe enters a mass spectrometer, either online (ESI-MS) or offline (MALDI-MS), where peptides are ionized to determine their mass over charge ratio (m/z), e.g. via a time-of-flight (TOF) analysis.

CP940, *CompLife 2007, 3rd International Symposium on Computational Life Science*
edited by A. Siebes, M. Berthold, R. Glen, and A. Feelders

With the help of the mass m, proteins can be identified by their mass fingerprints using database-driven computational tools like Mascot [1] or SEQUEST [2]. After these basic but essential analysis steps, more sophisticated investigation of the data, like e.g. quantification and comparative analysis, can be performed.

As in every scientific experiment, the quality of sample preparation and a correct calibration of the instruments in use significantly affect the quality of the output. Besides controllable factors, MS data is often disturbed by baseline and noise artifacts, hampering an efficient and exact interpretation of the measurements. In addition to electrical noise, mass spectra are often spiked with peaks caused by chemical impurities. This leads to a complex mixture of undesired contributions superimposed on the signal of interest. To overcome this problem, filtering techniques are often applied to ease the detection of peak-like structures by reducing the amount of baseline and noise artifacts. Finally, detected peaks are classified into 'noise' or 'signal' using local properties like their intensity or shape, or additional non-local features like the arrangement of mass peaks into characteristic isotopic distributions.

Following this common procedure, most algorithms in use today first apply a simple thresholding to separate between points of interest, so called *seeds*, and background noise. Obviously, the number of true and false positives strongly depends on the chosen threshold d. After the seeding step, each point of interest is extended to a suitable region in which a characteristic isotopic pattern should be matched. Regions that fail to fit a specific binomial distribution are classified as noise and discarded from further analysis.

Our contribution. In this work, we want to propose an alternative approach to the efficient and accurate detection of relevant features in complicated mass spectrometric data sets. We will describe a technique that uses additional knowledge about the signal's characteristics at the earliest possible stage, leading to greatly enhanced efficiency, while being astonishingly robust against both high- and low-frequent parasitics like spurious noise peaks and baseline terms. In principle, all available information about the signal of interest can be employed for the purposes of feature detection. On the other hand, an approach strongly adapted to the details of a given experiment will soon lose its generality. We thus decided to derive a wavelet based on general features of mass spectrometric data only. If desired, additional information can be easily integrated into the approach. The characteristics we chose to distinguish noise from signal are as follows: each molecular fragment we want to identify will give rise to a certain isotopic pattern in the mass spectrometric signal, composed of a number of individual peaks of a certain overall shape with a clearly defined spacing. Since peaks in the pattern may be lowly abundant and hard to detect, and due to the possibility of experimental deviations or overlapping signal contributions, this information can not be used as a hard filter, but rather as a soft criterion – regions in the mass spectrometric signal are scored by how well they fit with the above description. Fortunately, the facts described above already guarantee that the features of interest 'live' on a certain frequency band in between high-frequent noise and

low-frequent baseline terms (an isotopic pattern is wider than a typical noise peak but smaller than the length scales of baseline deviation). This motivates to construct a wavelet tailored to the shape of isotopic patterns and to detect features in a single scale of the corresponding wavelet transform. In addition to its ability to identify isotopic patterns instead of picking isolated peaks, the isotope wavelet transform provides information for an immediate determination of the charge state. For a single scan and a fixed length of the wavelet, the transform can be computed in linear time, making the isotope wavelet an accurate and fast technique for feature detection in mass spectrometric data. Consequently, a typical LC-MS data set consisting of several hundred scans can be analyzed in a few minutes on today's standard workstations. Besides its computational advantages, the method requires only a single parameter to be specified by the user. In our opinion, this simplicity is very important for real-world applications: often, advanced algorithms in mass spectrometry software involve the adaption of large numbers of parameters which are hard to interpret.
The wavelet described in this work has already been successfully employed in a myoglobin quantification study [3], where a replacement of the threshold-based seeding strategy described in [4] by the isotope wavelet greatly reduced the number of false positive seeds. Combined with the so called *sweep line paradigm*, the whole quantification ran about five times faster than before. Please note that in [3], the isotope wavelet has only been used, but not described or analyzed in detail. In this work, we give a full derivation of the isotope wavelet, an efficient algorithm for feature detection using the isotope wavelet transform, and an evaluation of the wavelet's results on selected MALDI data sets.

Related Work. The idea to use wavelets for analyzing mass spectrometry data is not new. Rejtar at al. [6] used wavelets to denoise mass spectra in order to improve database search. Instead of directly picking peaks within the wavelet transform they only filtered the data, requiring an additional preprocessing step. Du et al. [7] identified peaks by a series of continuous wavelet transforms with different scale parameters. While their approach overcomes background noise and baseline artifacts, its limiting factor is the computational load with about half a minute for a single scan. Lange et al. [5] used the Marr-wavelet for picking individual peaks, rather than complete isotopic patterns. While this allows for highly precise predictions of monoisotopic masses, the approach generates large numbers of false positives for each peak-like structure in the data – a property shared by all individual-peak based approaches. Interestingly, the high accuracy of individual-peak methods can be easily combined with the significantly better specificity of the isotope wavelet approach – as soon as regions of interest, i.e. isotopic patterns, have been identified in the signal, another peak picker can be employed to more accurately determine the monoisotopic masses of its peaks.

2 The Isotope Wavelet

As discussed above, the main ingredient to our feature detection scheme is a specifically tailored wavelet function. Fortunately, wavelet theory allows us to construct a function with the desired property of detecting isotopic patterns even in the presence of highly intensive parasitics. To motivate the construction process, the wavelet function can be thought of as a matched filter for the signal of interest, i.e. as a function featuring a high correlation with the pattern we are looking for, while fulfilling some (rather mild) criteria, depicted in the following.

2.1 The Continuous Wavelet Transform

The wavelet transform W_ψ of a function $f \in L^2(\mathbb{R})$ with respect to the wavelet ψ is given by

$$W_\psi[f](a,b) = \frac{1}{\sqrt{c_\psi\,|a|}} \int_{-\infty}^{\infty} f(t)\psi\left(\frac{t-b}{a}\right)\,dt, \qquad c_\psi := 2\pi \int_{-\infty}^{\infty} \frac{\left|\hat{\psi}(\omega)\right|^2}{|\omega|}\,d\omega \quad (1)$$

where $a \in \mathbb{R}\backslash\{0\}$ and $b \in \mathbb{R}$ [8]. a and b are called the scale and the translation parameter respectively and $\hat{\psi}(\omega)$ denotes the Fourier transform of $\psi(t)$. Please note that, as described above, for each charge z we will only need to search for features in a single scale of the wavelet transform. Thus, in the following we will only need to evaluate eq. (1) for one value of a per charge. Eq. (1) can be seen as a convolution of the signal with the translated and dilated wavelet. For reasons of simplicity, we will later use a correlation rather than a convolution for our isotope wavelet transform. This can be easily achieved, since convolution and correlation are linked by a mirrored wavelet function. Wavelets can be easily constructed using the following result:

Lemma 1. *If $0 \neq \psi \in L^2(\mathbb{R})$ has compact support and $\int \psi(t)\,dt = 0$, then ψ is a wavelet [8].*

2.2 The Isotope Wavelet

In order to derive a suitable filter, we will now describe how a wavelet with high correlation to isotopic patterns can be created. An isotopic pattern consists of a sequence of peaks, whose intensity ratios can be estimated from the mass of the fragment generating it. The individual peaks are evenly spaced at distances of m_n/z, where m_n denotes the mass of a neutron. Recent publications suggest to adapt m_n to slightly different values for statistical reasons (e.g. [9]).

To model the sequence of peaks, we start with a sine of a frequency compatible with the spacing of the pattern. The function is truncated after a certain number of oscillations to include only a sensible amount (e.g. 5) of peaks into the model. We multiply the function by a continuous equivalent of the Poisson distribution, yielding the correct intensity ratios for the maxima of the sine wave. Finally,

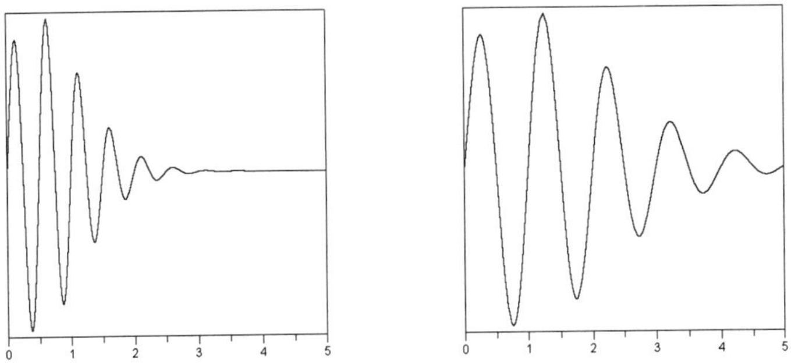

$m/z = 1000, z = 2$ $\qquad$ $m/z = 2000, z = 1$

Fig. 1. The isotope wavelet for different mass values and charges.

in order to construct a wavelet, we will need to subtract the function's mean value. The resulting function is obviously compactly supported and has – by construction – zero mean. Thus, according to Lemma 1, it is a wavelet. The positive parts of the wavelet closely resemble an isotopic pattern, while its negative parts 'punish' large spectral values in between peaks. Hence, large correlation of a signal with the isotope wavelet occurs if the signal resembles an isotopic pattern, while deviations from the perfect pattern structure will lead to a loss of correlation.

Formally, the isotope wavelet is defined as:

$$\Psi_{\lambda,z}(t) := H_0(t)\left(1 - H_\tau(t)\right)\left[\psi_{\lambda,z}(t) - \frac{1}{\tau}\int_0^\tau \psi_{\lambda,z}(t)\ dt\right] \tag{2}$$

$$\psi_{\lambda,z}(t) := \frac{\sin\left(\frac{2\pi z t}{m_n}\right) e^{-\lambda}\lambda^{tz}}{\Gamma(1+tz)} \tag{3}$$

H_τ denotes the Heaviside step function with $H_\tau(t) \neq 0$ within the interval $[\tau \ldots \infty)$ and zero outside. τ is a cutoff used to obtain compact support for Ψ in a trivial way. In addition, the isotope wavelet has two parameters: λ and z. While the latter one represents the charge state and therefore squeezes or stretches the wavelet function, respectively, λ controls the peak heights of Ψ. Since the intensity ratios between neighboring isotopic peaks change with m, the shape of the wavelet must be dynamically adapted during the correlation. Gay et al. [10] developed a polynomial model resembling the mean mass signals within isotopic patterns. Starting with the monoisotopic peak they provide six polynomial functions approximating the height of each peak triggered by an averagine peptide. Hence, we were able to fit $\lambda(m)$ – and therefore to adapt the wavelet's shape – by linear regression. Figure 1 shows two instances of the isotope wavelet for different mass values and charge states.

Pseudo-Code for the Isotope Wavelet Transform

Let S denote a set of mass spectra and $Z := \{1, \ldots, z_{max}\}$ a set of charges. For some signal s let $\overline{s}$ denote its average value and $\hat{\sigma}(s)$ an estimate of its standard deviation.

1. `Compute Isotope Wavelet Transform`

   $\forall\ s \in S, z \in Z:$ `compute` $W_{\Psi_{\lambda,z}}[s(t)](p_i) := s \star \Psi_{\lambda(t_i),z}$
   with p_i as the i-th m/z position
   and $\star$ denoting the correlation operator

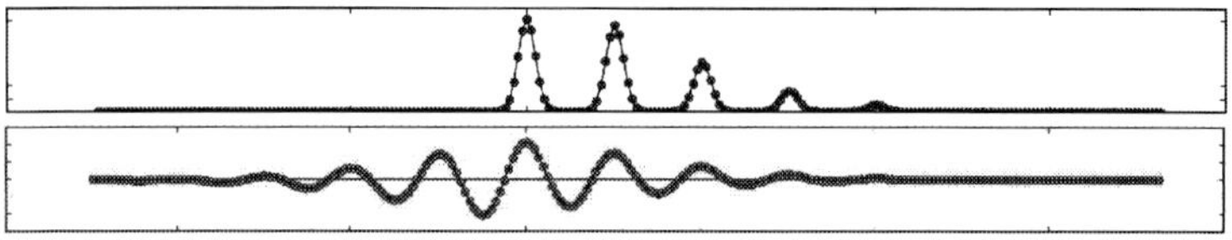

Fig. 2. A spectrum s and its corresponding charge-1-transform.

2. `Filtering`

   $\forall\ w \in \{W_{\Psi_{\lambda,z}}[s] \mid s \in S\}:$ `sort` w `by decreasing amplitude`
   `drop all` $p_i \in w$ `with` $w(p_i) \leq d$
   for $d := \tilde{d}\hat{\sigma}(w) + \overline{w}$
   and $\tilde{d} \geq 0$ as a threshold parameter

   Let $F_{\Psi_{\lambda,z}}[s]$ denote the filtered and sorted versions of $W_{\Psi_{\lambda,z}}[s]$.

3. `Scoring`

   $\forall\ f \in \{F_{\Psi_{\lambda,z}}[s] \mid s \in S\}: \ \forall\ p_i \in f:$
   `compute` $P_{\lambda,z}[s](p_i) := \sum_{j=1-\tau}^{\tau-1} f\left(p_i + m_n \frac{j}{z}\right) - \sum_{j=1-\tau}^{\tau-2} f\left(p_i + m_n \frac{j+\frac{1}{2}}{z}\right)$

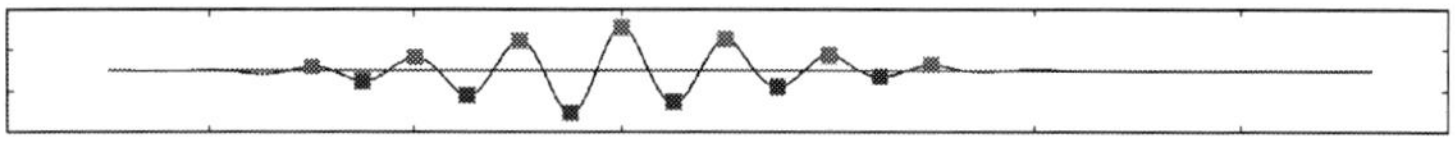

Fig. 3. Scoring by expected local maxima and minima.

4. `Isotope Picking / Charge Prediction`

   $\forall\ f \in \{F_{\Psi_{\lambda,z}}[s] \mid s \in S\}: \ \forall\ p_i \in f:$
   `if` $P_{\lambda,z}[s](p_i) \geq d + f(p_i)$
   `assign charge` z_m `to` p_i `with` $z_m := \text{argmax}_z \{P_{\lambda,z}[s](p_i)\}$

2.3 Picking Isotopic Patterns

While Ψ can be adapted to the mass m on the fly, i.e. while we shift the wavelet function over the signal, this is obviously not possible for the parameter z. Hence, for each scan s contained in the LC-MS map, we have to compute several wavelet transforms $W_{\Psi_{\lambda,z}}[s(t)](a,b)$ for $z = 1 \dots z_{max}$.
Having computed the isotope wavelet transform, every point with an amplitude exceeding some threshold parameter d can be considered as a potential seed, as it features positive correlation of experimental signal and isotope wavelet. Unfortunately, it often turns out insufficient to base the classification of a single point into either 'noise' or 'signal' exclusively on the amplitude. Even isolated peaks with large intensity trigger a response in the transform exceeding the threshold parameter d. Hence, we established a more advanced scheme to distinguish between true and false positives. Consider Figure 2 (within the pseudo-code), where an artificial mass spectrum s with ideal isotopic pattern of charge state 1 is depicted. Its corresponding charge-1-transform $W_{\Psi_{\lambda,1}}[s]$ shows a characteristic wave-like structure also containing negative intensities, where signal and wavelet are anti-correlated. We exploit the typical alternating sequence of valleys and peaks and sum the absolute intensity values at these local extrema to form a final score $P_{\lambda,z}[s](p_i)$ for every point p_i (c.f. pseudo-code).

3 Results

Since the isotope wavelet has already been successfully applied to an LC-ESI data set in the context of a myoglobin quantification study [3], we will restrict our analysis in this work to matrix-assisted laser desorption/ionization (MALDI) data. To this end, we will discuss our results on the data sets BSA-1, BSA-2 and BGAL: while both BSA data sets contain a tryptic digest of bovine serum albumin, data set BSA-1 has been measured after reduction of disulfids and protection of cystein residues by iodoacetamide. BSA-2 has been recorded on a modern Applied Biosystems machine (ABI 4800 TOF/TOF) with excellent signal-to-noise behavior, while BSA-1 has been measured on an older machine and is consequently much worse resolved. The last data set, BGAL, contained a tryptic digest of β-galactosidase and has also been acquired on the ABI machine.

Dataset BSA-1 has been previously used in [5] to compare peak picking accuracy based on a Mascot [1] peptide mass fingerprinting. Using the peaks listed by the vendor (Bruker) software, a sequence coverage of 44% has been reported, while the Marr wavelet by Lange et al. achieved coverages between 52% and 67% depending on the threshold value. In our experiments, the isotope wavelet yielded up to 69% coverage. Since the analysis of peak picking performance in [5] was focused on mass accuracy, not false positive rate, no information is provided about the number of features detected in total.

In order to demonstrate that individual-peak picking techniques usually lead to large numbers of false positive hits, we compared the isotope wavelet to the Decon2LS software tool [11]. Table 3 demonstrates that algorithms picking peaks only based on a minimum signal-to-noise ratio (S/N) can – by construction – not

Vendor Software				
S/N	BSA	Features	Score	Coverage
0	✓	116	**150**	0.43
1	✓	116	**150**	0.43
3	✓	116	136	**0.44**
5	✓	53	102	0.28
7	✓	52	103	0.28
10	✓	52	103	0.28
15	✓	35	124	0.24
20	✓	34	126	0.24
30	✓	30	105	0.23
50	✓	20	88	0.17
100	✓	13	82	0.14

Isotope Wavelet				
$\tilde{d}$	BSA	Features	Score	Coverage
0	✓	213	**222**	**0.56**
1	✓	196	177	0.55
2	✓	121	179	0.46
3	✓	90	162	0.4
4	✓	78	130	0.35
5	✓	61	119	0.3
7	✓	55	126	0.26
9	✓	37	122	0.25
10	✓	35	106	0.23
11	✓	30	97	0.21
15	✓	22	83	0.15

Decon2LS Peak Finder				
S/N	BSA	Features	Score	Coverage
0	*n.a.*	3443	*n.a.*	*n.a.*
1	*n.a.*	3443	*n.a.*	*n.a.*
3	*n.a.*	2963	*n.a.*	*n.a.*
5	*n.a.*	2478	*n.a.*	*n.a.*
7	*n.a.*	2223	*n.a.*	*n.a.*
10	*n.a.*	1718	*n.a.*	*n.a.*
15	*n.a.*	1339	*n.a.*	*n.a.*
20	*n.a.*	1214	*n.a.*	*n.a.*
30	*n.a.*	1106	*n.a.*	*n.a.*
50	*n.a.*	1062	*n.a.*	*n.a.*
100	*n.a.*	1031	*n.a.*	*n.a.*
110	–	35	*n.a.*	*n.a.*
120	✓	33	66	0.19

Decon2LS Mass Transform				
S/N	BSA	Features	Score	Coverage
0	✓	214	70	0.3
1	✓	214	81	0.3
3	✓	234	78	**0.32**
5	✓	197	79	**0.32**
7	✓	181	**84**	0.28
10	✓	140	70	0.23
15	✓	100	71	0.22
20	✓	85	58	0.18
30	–	66	*n.a.*	*n.a.*
50	–	56	*n.a.*	*n.a.*
100	–	48	*n.a.*	*n.a.*
110	–	10	*n.a.*	*n.a.*
120	–	10	*n.a.*	*n.a.*

Table 1. Results for BSA-2.

The column 'BSA' indicates if Mascot identified bovine serum albumin as the probe's ingredient. While 'Score' and 'Coverage' also relate to the Mascot results, the column named 'Features' lists the number of peaks identified by the corresponding software and parameter settings. Sometimes Mascot scores were not available due to the hugh number of identified features.

discriminate isotopic patterns from peaked noise, in particular if the thresholding parameter is small. On the other hand, raising the S/N ratio simultaneously decreases the sequence coverage, since the peaks generated by low abundant or weakly ionizing peptides will be lost. Hence, for a minimum S/N of 3, the Decon2LS software identified nearly 3,000 peaks in a spectrum consisting of 'only' 150,000 data points. The resulting number of peaks can be substantially reduced using the algorithm proposed by Horn et al. [9], also implemented in the Decon2LS software package. After this filtering step, the best results for Mascot score and sequence coverage for this technique were given by 84 and 32% respectively, while for some S/N settings, identification even failed completely.

On the same data set, we tested the performance of the vendor (ABI) software supplied with the mass spectrometer for different S/N settings, and finally our isotope wavelet method for different values of $\tilde{d}$. Here, the vendor software yielded a maximum sequence coverage of 44% for a S/N of 4, while the best Mascot score of 150 was found for a S/N of 1. The isotope wavelet, in contrast, achieves a nicely improved coverage of 56% and a significantly higher score of 222.

The last MALDI data set contained a tryptic digest of β-galactosidase. While for the other two data sets, reference information on Mascot fingerprints using other peak picking schemes was available, for BGAL we can only report the main results for the isotope wavelet: β-galactosidase has been identified with a score of 303 and a coverage of 49%, again demonstrating the good performance of our approach as already observed on the other data sets.

4 Conclusions and Outlook

In the presented work, we have described a new technique for the accurate, sensitive, and specific detection of isotopic features in complex mass spectrometric data sets. The method is sufficiently rapid to be applied to the large data sets occurring in real-world applications. The main ingredient of our method is a wavelet function tailored to the detection of isotopic patterns, and an algorithmic scheme for the scoring of potential patterns based on their wavelet transform.

The method has been tested successfully on a number of different data sets, where it has consistently produced very high Mascot scores for the correct identification with large sequence coverage and a comparably low number of false positives. Particularly important in our opinion is the fact that the whole approach presented in this work requires only a single parameter to be set by the user – a threshold that can be used to shift the sensitivity versus specificity behavior of the method. At least for single MS scans, our experiments have shown that typically, we can set this parameter to 0 and only need to increase the value if the number of detected features becomes too large for the application at hand. Hence, the method can be used 'out of the box', without the need to identify suitable values for a large number of parameters and optimizing them for each given set of experimental data, as it is often the case in programs for proteomics data analysis. In summary, we were able to show that the inclusion of non-local information about mass spectrometric data – in our case the shape of isotopic distributions – can lead to significantly improved feature detection, while still being computationally highly efficient. The wavelet-based approach taken in this work is general enough to allow for the inclusion of more detailed information about the spectra at hand, e.g. about typical patterns arising for labeled mass spectrometric data. Currently, work is under way to identify such information and to include it into the wavelet design process, leading to tailored wavelets for different kinds of mass spectrometric experiments.

Acknowledgements

The authors would like to express their thanks to Johan Gobom (Free University Berlin) who provided the BSA-1 data set and to Marc Sturm for his support in using the OpenMS library and ToppView [12]. Rene Hussong and Andreas Hildebrandt have been supported by DFG grants BIZ4:1-4.

Appendix

All Mascot queries have been performed with a peptide mass tolerance of 100 ppm (only exception – BSA-1: 150 ppm), a maximum of one missed cleavage site and 'Oxidation(M)' as variable modification. Parameters for the ABI software were as follows: Local Noise Width (m/z) 250, Min Peak Width at FWHM 2.9. The Cluster Area Optimization S/N threshold has been dynamically adapted to about three times the S/N threshold as suggested by ABI's documentation. Decon2LS's parameters have been left unchanged to the preset values.

References

1. D.N. Perkins, D.J.C. Pappin, D.M. Creasy, J.S. Cottrell: Probability-based protein identification by searching sequence databases using mass spectrometry data. Electrophoresis **20** (1999) 3551–3567.
2. J.K. Eng, A.L. McCormack, J.R. Yates III: An Approach to Correlate Tandem Mass Spectral Data of Peptides with Amino Acid Sequences in a Protein Database. J Am Soc Mass Spectrom **5** (1994) 976–989.
3. O. Schulz-Trieglaff, R. Hussong, C. Gröpel, A. Hildebrandt, K. Reinert: A Fast and Accurate Algorithm for the Quantification of Peptides from Mass Spectrometry Data. Proceedings of the 11th RECOMB, LNBI 4453 (2007) 473–487.
4. C. Gröpl, E. Lange, K. Reinert, O. Kohlbacher, M. Sturm, C.G. Huber, B.M. Mayr, C.L. Klein: Algorithms for the automated absolute quantification of diagnostic markers in complex proteomics samples. Proceedings of the 1st CompLife (2005) 151–163.
5. E. Lange, C. Gröpl, K. Reinert, O. Kohlbacher, A. Hildebrandt: High-accuracy peak picking of proteomics data using wavelet techniques. Pac. Symp. Biocomput. (2006) 243–254.
6. T. Rejtar, H.S. Chen, V. Andreev, E. Moskovets, B.L. Karger: Increased Identification of Peptides by Enhanced Data Processing of High-Resolution MALDI TOF/TOF Mass Spectra Prior to Database Searching. Anal. Chem. **76** (2004) 6017–6028.

7. P. Du, W.A. Kibbe, S.M. Lin: Improved peak detection in mass spectrum by incorporating continuous wavelet transform-based pattern matching. Bioinformatics **22** (2006) 2059–2065.
8. A.K. Louis, P. Maaß, A. Rieder: Wavelets: Theory and Applications. John Wiley & Sons (1997).
9. D.M. Horn, R.A. Zubarev, F.W. MacLafferty: Automated reduction and interpretation of high resolution electrospray mass spectra of large molecules. J Am Soc Mass Spectrom **11** (2000) 320–332.
10. S. Gay, P.-A. Binz, D.F. Hochstrasser, R.D. Appel: Modeling peptide mass fingerprinting data using the atomic composition of peptides. Electrophoresis **20** (1999) 3527–3534.
11. N. Jaitly, M.E. Monroe, N.Tolić, K. Littlefield, D. S. Daly, J.N. Adkins, G.A. Anderson, R. D. Smith: Open Source Tools for the Accurate Mass and Time (AMT) Tag Proteomics Pipeline. Poster presentation at ASMS (2006).
12. O. Kohlbacher, K. Reinert, C. Gröpl, E. Lange, N. Pfeifer, O. Schulz-Trieglaff, M. Sturm: TOPP–the OpenMS proteomics pipeline. Bioinformatics **23** (2007) 191–197.

Calculating Mean First Passage Times from Markov Models of Proteins

Christian H. Jensen, Dmitry Nerukh and Robert C. Glen

Unilever Centre for Molecular Science Informatics, Department of Chemistry, University of Cambridge, Cambridge CB2 1EW, UK

Abstract. Assuming that the dynamics of peptides and proteins can be described by Markov transitions between different configurational states, we present a method which can calculate the mean first passage time (MFPT) between *sets* of initial *and final* states. The method is described in detail and differences between this and a method commonly employed [1] are explained. It is shown that the proposed method is (i) more general in allowing sets of final states, (ii) significantly faster, (iii) more accurate since it does not involve the calculation of infinite summations. Particular attention is given to the biologically important case of multiple final states.

1 Introduction

There are many methods which seek to simulate the folding of a peptide or protein. They range from the very course grained approaches like the HP model [2] to models with atomic detail like the Molecular Dynamics approach [3]. While the course grained method gives results which can be useful as guidelines when designing proteins, they do not describe exactly how a specific protein folds. To describe how a specific protein in atomistic detail folds, a model like Molecular Dynamics is needed. The problem with the latter is that for system sizes of bio-chemical interest the computational task of making a single Molecular Dynamics simulation which shows a complete folding process is infeasible. However, recently a method which combines several Molecular Dynamics simulations by using clustering and a Markov model for the state transitions has been proposed [1]. Using this method it is possible to reconstruct the overall dynamics of a peptide from thousands of individual simulations. This can be done by first clustering the configurational space into discrete states, and then count the number of transitions between the different states from all the simulations. The Markov model can be described by a state vector v and a transition matrix T. Given that the system is described by a state vector v_t at time t the state vector at time $t+1$ can be calculated as $v_{t+1} = Tv_t$.

One of the fundamental problems that arises in the analysis of such simulations is the definition of the average folding times in cases when several states assigned by the algorithm can be considered as "final". From the bio-chemical point of view the native states of proteins are defined by their ability to perform biological functions. The latter is often assigned to a "functional core" of

CP940, *CompLife 2007, 3rd International Symposium on Computational Life Science*
edited by A. Siebes, M. Berthold, R. Glen, and A. Feelders

the molecule while the rest of the molecule is allowed to "breath" producing configurations that differ sometimes very significantly [4]. Clustering algorithms produce many states described above and formal coarse graining can not be used here because of the high risk of loosing important details in the functionally relevant parts of the molecule. Thus, the most natural solution of the problem lies in assigning a set of final states rather than trying to artificially lump states together.

In this paper, we present a method for calculating the average folding time from a Markov model. This method generalises to calculating the folding time between multiple initial and final states. This is different from the method [1] previously and most commonly used (see, for example, [5, 6]) and we investigate cases where multiple final states lead to significant differences in the folding times. For degenerate cases, the differences turn out to be critical, however, the differences are also shown to be very substantial (up to ~70%) for a Markov model constructed from a Molecular Dynamics simulation.

2 Theory

To calculate the average transition time of a Markov model we need to define initial and final states. Each of these can either be one state or a set of states. Assuming that we have a set of initial states I and a set of final states F the average transition time can be written as:

$$t_{IF} = \sum_{n=1}^{\infty} n P_{IF}(n) \tag{1}$$

Here $P_{IF}(n)$ is the probability for all paths of length n which start in I and end on F. We assume that the Markov process is described by a transition matrix T and that there is a total of N states. The first problem in the calculation is to find an expression for $P_{IF}(n)$. Below we introduce $\tilde{T}$,L,o and v.

- From the transition matrix T remove the rows and columns for all states in F to form a new matrix $\tilde{T}$. This new matrix will have a dimension of $(N-d) \times (N-d)$, where d is the number of states in F.
- Form matrix L which is of dimension $d \times (N-d)$ and holds the matrix elements of T that give the probabilities for entering F from all other states.
- Form row vector o which is of dimension $1 \times d$ and holds ones in all places.
- Form vector v of dimension $(N-d) \times 1$. The elements of v must describe the initial distribution of states in I. If each starting state is equally likely then their elements must be equal. For the states not in I the initial value in v must be zero. The total sum of all elements in v must be 1.

Using the quantities given above $P_{IF}(n)$ can be written as $oL\tilde{T}^{n-1}v$ (an explanation for this is given in Appendix A). Let us assume that $\tilde{T}$ has eigenvectors

e_i with corresponding eigenvalues λ_i. We then expand v in this basis. This gives $v = \sum_i \alpha_i e_i$. The average transition time (1) can then be written as:

$$t_{IF} = \sum_{n=1}^{\infty} nP_{IF}(n) = \sum_{n=1}^{\infty} noL\tilde{T}^{n-1}v \tag{2}$$

$$= \sum_{n=1}^{\infty} noL\tilde{T}^{n-1} \sum_i \alpha_i e_i = \sum_{n=1}^{\infty} \sum_i noL\alpha_i \lambda_i^{n-1} e_i \tag{3}$$

$$= \sum_i \left(\sum_{n=1}^{\infty} n\lambda_i^{n-1} \right) \alpha_i oLe_i = \sum_i \frac{\alpha_i}{(1-\lambda_i)^2} oLe_i \tag{4}$$

The method proposed differs from the commonly used method [1] in (i) it allows sets of *final* states, while the latter one only permits a single final state, (ii) it does not involve the calculation of infinite summations, thus reducing the calculation time significantly, (iii) for the same reason our method is more accurate since the calculation of the eigenvectors and eigenvalues is a well developed area of applied mathematics.

3 Results

In this section we apply the method described to three examples. Firstly we investigate two artificial degenerate cases which describe extreme folding situations. One case is when only one folding path exists and the other extreme is when transitions between all states exist so that all paths are possible. Thereafter we investigate how the computational time scales with the number of states in a given transition matrix.

The situation with only one folding path can be described by the transition matrix (5). There are a total of five states in this matrix. Labeling with numbers from 1 to 5 we consider a situation where the initial state is 1 and the set of final state is 4,5. Using the method given in the literature [1] the average folding times from state 1 to 4 and from state 1 to 5 can be found. They can then be compared with the time of reaching the set of states 4 and 5. The average folding time from state 1 to 4 is 45.0 time steps and the average folding time from state 1 to 5 is 80.0 time steps. Thus, it is critical which state to consider as a "final" one even though they are the neighbouring states on the folding path.

Using the method proposed in this work, the average folding from state 1 to the set of states 4 and 5 is found to be 45.0 time steps. This is the same as the average folding time between states 1 and 4. The reason why this result is correct is that there is only one folding path between states 1 and 5. This means that it is not possible to reach state 5 from state 1 without first passing through state 4. Hence, only the time taken to arrive to state 4 which is of importance defines the folding time to the set 4,5.

$$\begin{bmatrix} 0.8000 & 0.1000 & 0.0000 & 0.0000 & 0.0000 \\ 0.2000 & 0.8000 & 0.1000 & 0.0000 & 0.0000 \\ 0.0000 & 0.1000 & 0.8000 & 0.1000 & 0.0000 \\ 0.0000 & 0.0000 & 0.1000 & 0.8000 & 0.2000 \\ 0.0000 & 0.0000 & 0.0000 & 0.1000 & 0.8000 \end{bmatrix} \quad (5)$$

The other limiting case, the situation with multiple folding paths can be represented by the transition matrix (6). Again there is a total of five states in the matrix and we take state 1 to be the initial state and states 4 and 5 to be the final states. By using the method from [1] the calculated average folding time from state 1 to 4 is 20.0 time steps and from state 1 to 5 is also 20.0 time steps. Our method, however, results in 10.0 time steps for the 4,5 set. The reason why this result is correct is because compared to a single final state the probability of entering two final states from all other states is doubled. Since this probability is doubled in each time step the average folding time is halved.

$$\begin{bmatrix} 0.8000 & 0.0500 & 0.0500 & 0.0500 & 0.0500 \\ 0.0500 & 0.8000 & 0.0500 & 0.0500 & 0.0500 \\ 0.0500 & 0.0500 & 0.8000 & 0.0500 & 0.0500 \\ 0.0500 & 0.0500 & 0.0500 & 0.8000 & 0.0500 \\ 0.0500 & 0.0500 & 0.0500 & 0.0500 & 0.8000 \end{bmatrix} \quad (6)$$

Thus, in both limiting cases it is fundamentally important to consider multiple final states as *sets* of states. Considering them individually leads to significantly different results in the mean passages times.

The cases discussed above are constructed examples. To test a realistic situation a transition matrix was obtained from a Molecular Dynamics simulation of the four residue peptide VPAL solvated in 874 water molecules. The simulation box was 3.0x3.0x3.0Å. The system was equilibrated before it was sampled for 500ns. The force field was 53a6 [7, 8, 9] and the temperature was held at 300K during the simulation. The Molecular Dynamics trajectory was then clustered in configurational space and the number of transitions between these states was counted. From this the transition matrix (7) was calculated.

As in the previous cases we take state 1 to be the initial state and states 4 and 5 to be the final states. Following the method from [1] the average folding time from state 1 to 4 was calculated to be 277.7 ps and 384.7 ps for the 1 to 5 transition. The average folding time from state 1 to the set 4,5 obtained with our method is 220.0 ps. The difference demonstrates that not only limiting cases can produce substantial differences in calculated average folding times.

$$\begin{bmatrix} 0.9940 & 0.0080 & 0.0018 & 0.0220 & 0.0001 \\ 0.0023 & 0.9802 & 0.0000 & 0.0000 & 0.0232 \\ 0.0016 & 0.0003 & 0.9973 & 0.0073 & 0.0006 \\ 0.0020 & 0.0000 & 0.0007 & 0.9505 & 0.0152 \\ 0.0000 & 0.0115 & 0.0000 & 0.0201 & 0.9610 \end{bmatrix} \quad (7)$$

From the matrix we can see that it is not of the form of either of the matrices 5 or 6. The new matrix can be described as a combination of the two. Therefore, the difference in average folding time can also be understood by a combination of a limited number of folding paths (5) and an increase in the probability at each time step for reaching the final states (6). For larger peptide and protein systems possibly involving thousands of states the difference in the calculated average folding times may be less significant, however, for a given example this is difficult to quantify without calculating the correct average folding time given by Equation 1.

To investigate how the computational time of the algorithm scales with the number of states in the transition matrix we have done simulations with transition matrices of different size. The transition matrices are chosen so that after each time step there is an 80% chance of staying in the same state and there is an even probability of entering all other states. The transition matrix is therefore of the same form as the transition matrix (6). As the transition matrix is symmetric any choice of initial and finial states will give the same results.

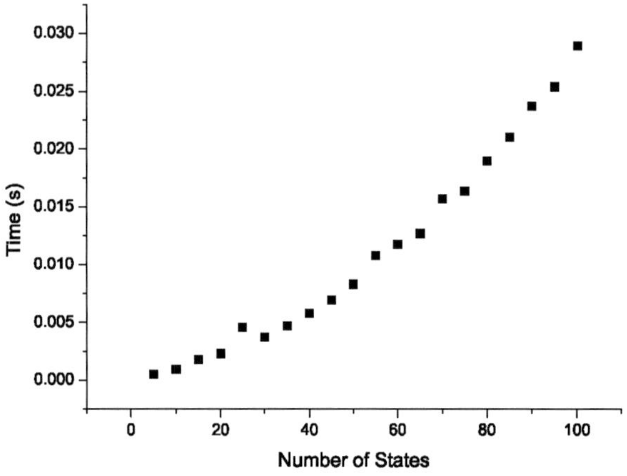

Fig. 1. Running time against the number of states in the transition matrix for the new algorithm.

The results can be seen in Figure 1. It is found that the running time of the algorithm increases with the number of states in the transition matrix. In the algorithm most time is spent calculating the eigenvectors. Calculating eigenvectors scales as $O(n^3)$. For large n this behavior will therefore describe the running time of the algorithm. Once the eigenvectors are found calculating the MFPT is done easily by Equation 2 and given that the eigenvectors are calculated accurately the MFPT will also be accurate. In Figure 2 the equivalent results are shown for a method which involves computing an infinite sum. For this method the running time depends on the accuracy needed. Here we have chosen a cutoff of 0.0001. This is accurate enough for transition matrices with few states. However, as the number of states increases the cutoff will have to be decreased to maintain the accuracy. Therefore the linear behavior in Figure 2 is artificial. It is also noted that even with the modest accuracy chosen the new algorithm are orders of magnitude faster.

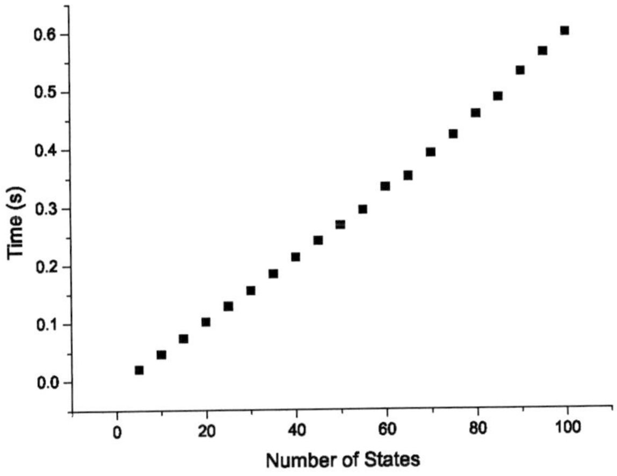

Fig. 2. Running time against the number of states in the transition matrix for a method which involves an infinite summation.

4 Conclusions

In this work we present an efficient method for obtaining the average folding time between sets of states. By applying the method to two constructed degenerate transition matrices we demonstrate that the results are fundamentally different from the individual transitions to the states in the final set. The method was also applied to a transition matrix of a four residue peptide obtained from a Molecular Dynamics simulation. Our method predicted an average folding time of 220.0 ps to a set of two states whereas the commonly used method from [1] resulted in 277.7 and 384.7 ps for the states individually. It should be stressed that taking the average of the times for the individual states is incorrect. The only solution in this situation would be joining together all final states by forming a single one and then applying the method [1]. This, however, would involve recalculation of the transition matrix and that would result in significant computational overheads for realistic molecules having thousands of states.

5 Acknowledgments

The work is supported by Unilever and the European Commission (EC Contract Number 012835 - EMBIO). The authors thank Dr Maxim Fedorov for useful comments on the interpretation of the results.

References

1. Jayachandran, G. and Vishal, V. and Pande, V. S.: Using massively parallel simulation and Markovian models to study protein folding: Examining the dynamics of the villin headpiece. Journal of Chemical Physics. **124** 16 (2006) 164902
2. Lau, K. F. and Dill, K. A.: A lattice statistical-mechanics model of the conformational and sequence-spaces of proteins. Macromolecules. **22** 10 (1989) 3986-3997
3. Adcock, S. A. and McCammon, J. A.: Molecular dynamics: Survey of methods for simulating the activity of proteins. Chemical Reviews. **106** 5 (2006) 1589-1615
4. Alexei V. Finkelstein and Oleg Ptitsyn: Protein Physics, A Course of Lectures. Academic Press. (2002)
5. Krivov, Sergei V. and Karplus, Martin: Hidden complexity of free energy surfaces for peptide (protein) folding. PNAS. **101** 41 (2004) 14766-14770
6. Kasson, Peter M. and Kelley, Nicholas W. and Singhal, Nina and Vrljic, Marija and Brunger, Axel T. and Pande, Vijay S.: Ensemble molecular dynamics yields submillisecond kinetics and intermediates of membrane fusion. PNAS. **103** 32 (2006) 11916-11921
7. Hess, B. and van der Vegt, N. F. A.: Hydration thermodynamic properties of amino acid analogues: A systematic comparison of biomolecular force fields and water models. Journal of Physical Chemistry B. **110** 35 (2006) 17616-17626
8. Oostenbrink, C. and Soares, T. A. and van der Vegt, N. F. A. and van Gunsteren, W. F.: Validation of the 53A6 GROMOS force field. European Biophysics Journal with Biophysics Letters. **34** 4 (2005) 273-284
9. Oostenbrink, C. and Villa, A. and Mark, A. E. and Van Gunsteren, W. F.: A biomolecular force field based on the free enthalpy of hydration and solvation: The GROMOS force-field parameter sets 53A5 and 53A6. Journal of Computational Chemistry. **25** 13 (2004) 1656-1676

A Calculation of $P_{IF}(n)$

To illustrate how $P_{IF}(n)$ is calculated let us consider a three state system. Let the initial state be 1 and the final state 3. The transition matrix for the system is given as:

$$T = \begin{bmatrix} a_{11} & a_{12} & a_{13} \\ a_{21} & a_{22} & a_{23} \\ a_{31} & a_{32} & a_{33} \end{bmatrix} \tag{8}$$

First we form the matrices $\tilde{T}$, L, o and v:

$$\tilde{T} = \begin{bmatrix} a_{11} & a_{12} \\ a_{21} & a_{22} \end{bmatrix}, \quad L = \begin{bmatrix} a_{31} & a_{32} \end{bmatrix}, \quad o = \begin{bmatrix} 1 \end{bmatrix} \quad v = \begin{bmatrix} 1 \\ 0 \end{bmatrix} \tag{9}$$

For $n = 1$ we have:

$$P_{13}(1) = oL\tilde{T}^0 v = a_{31} \tag{10}$$

Since $P_{13}(1)$ is the probability to go from state 1 to state 3 in one step there is only one possible path, 1-3. The probability for this is simply a_{31}. For $n = 2$ we have:

$$P_{13}(2) = oL\tilde{T}^1 v = a_{31}a_{11} + a_{32}a_{21} \tag{11}$$

There are two possible paths 1-1-3 and 1-2-3. The probability for each of these is $a_{31}a_{11}$ and $a_{32}a_{21}$ respectively. The sum, therefore, gives the total probability. For $n = 3$ we have:

$$P_{13}(3) = oL\tilde{T}^2 v = a_{31}a_{11}a_{11} + a_{31}a_{12}a_{21} + a_{32}a_{21}a_{11} + a_{32}a_{22}a_{21} \tag{12}$$

In this case there are four possible paths from state 1 to 3. These are 1-1-1-3, 1-2-1-3, 1-1-2-3 and 1-2-2-3. $P_{13}(3)$ is the sum of the probabilities for each of these paths.

Predicting Error Bars for QSAR Models

Timon Schroeter[1,2], Anton Schwaighofer[1], Sebastian Mika[3], Antonius Ter Laak[4], Detlev Suelzle[4], Ursula Ganzer[4], Nikolaus Heinrich[4], and Klaus-Robert Müller[2,1]

[1] Fraunhofer FIRST, Kekuléstraße 7, 12489 Berlin, Germany
timon.schroeter@first.fraunhofer.de
[2] Technische Universität Berlin, Department of Computer Science, Franklinstraße 28/29, 10587 Berlin, Germany
[3] idalab GmbH, Sophienstraße 24, 10178 Berlin, Germany
[4] Research Laboratories of Bayer Schering Pharma AG, Müllerstraße 178, 13342 Berlin, Germany

Abstract. Unfavorable physicochemical properties often cause drug failures. It is therefore important to take lipophilicity and water solubility into account early on in lead discovery. This study presents $\log D_7$ models built using Gaussian Process regression, Support Vector Machines, decision trees and ridge regression algorithms based on 14556 drug discovery compounds of Bayer Schering Pharma. A blind test was conducted using 7013 new measurements from the last months. We also present independent evaluations using public data. Apart from accuracy, we discuss the quality of error bars that can be computed by Gaussian Process models, and ensemble and distance based techniques for the other modelling approaches.

1 Introduction

A large fraction of drug failures can be attributed to an unfavorable PC-ADME/T profile[5] [1,2]. Aiming at avoiding late (i.e., expensive) failures, many pharmaceutical companies have started to consider these properties, including log P and log D, already in the lead discovery phase.

Yet, makers of commercial predictive tools are often not able to incorporate data of the kind that is typically investigated in pharmaceutical companies. The tools are therefore often constructed using publicly available measurements of relatively small and mostly neutral molecules. Often, their accuracy on the in-house compounds of pharmaceutical companies is relatively low [3]. In our work, we use machine learning tools to construct models that are tailored to the specific classes of compounds that are investigated in drug discovery (in our case, in the drug discovery process at Bayer Schering Pharma). We employ a descriptor based modelling approach, therefore structures do not need to be disclosed to the modelling team (in our case, a research group that is not part of Bayer Schering Pharma). Information about our initial work on building such

[5] physico chemistry, absorption, distribution, metabolism, excretion, toxicity

CP940, *CompLife 2007, 3rd International Symposium on Computational Life Science*
edited by A. Siebes, M. Berthold, R. Glen, and A. Feelders

customized models to predict lipophilicity can be found in [4], for background information see [5].

Our experiments showed that we can achieve highest accuracy with so-called Gaussian Process (GP) models [6], a Bayesian nonlinear regression method. The Bayesian framework provides theoretically well founded criteria for model selection, i.e., choosing the "right amount of nonlinearity". This allows in turn for a fully automatic model building. Thus, the user of these models (Bayer Schering Pharma) can easily repeat the model building once new experimental data becomes available, in order to generate an improved model.

As a further major advantage, Gaussian Process models can provide an error bar for each individual prediction. This error bar can be used to distinguish predictions made inside the domain of applicability (small error bar) from less reliable predictions outside the domain of applicability (large error bar). In contrast to commonly used error bar heuristics [7], GP error bars are an integral part of model building and have a solid statistical foundation.

2 Methods and Data

2.1 Data Preparation

Dataset 1 contains 14556 drug discovery molecules of Bayer Schering Pharma. log D was measured at $pH = 7.0$ through 7.4. Although for particular compounds with pKa-values close to $pH = 7$ one can expect deviations in logD of up to 0.4 (extreme case), preliminary experiments showed that no negative impact on the model accuracy when collecting these measurements in a single data set. *Dataset 2* (in-house validation) is a set of 7013 new measurements of drug discovery molecules of Bayer Schering Pharma from the last months, including compounds from new projects. *Dataset 3* (public) consists of measurements of log P for 7926 unique compounds from the Physprop [8] and Beilstein [9] databases. To avoid log D values wrongly reported as log P values, the set was restricted to compounds predicted to be completely neutral at pH 2 to 11 by ACDLabs v9, since for these compounds, log D values in the given pH ranges coincide with the correct log P values. The compound identifiers of dataset 3 are included in the supporting information to our initial work on lipophilicity [4] and can be downloaded from `http://chemmedchem.org`. For more detailed information on how we pre-processed the data see [5].

For each of the molecules, we use Corina [10] to generate a 3D structure, and subsequently compute molecular descriptors using the Dragon [11] software. These descriptors are then used as features for different machine learning algorithms, with the goal of inferring the relationship between the descriptors and the experimental values for log P and log D_7.

The median of molecular weight is 227 g/mol for the public dataset, 432 g/mol for the in-house set and 405 g/mol for the in-house validation set. More than 90 % of the compounds in the public set have a molecular mass lower than 400 g/mol. Care has to be taken when evaluating models on mixed sets, since such models typically perform well on compounds with low molecular weight but are less accurate for the larger compounds relevant to drug discovery. On the *in-house* and *public* set of data, models are evaluated in 2-fold cross-validation ("leave 50 % out cross-validation") that is repeated 10 times [12]. The final Gaussian Process models built by the modelers at Fraunhofer FIRST and idalab on the *in-house* set of data were evaluated by researchers at Bayer Schering Pharma on the *in-house validation* set of data. At this point in time, the modelers had no knowledge of the nature or log D values of the validation set. Later, the validation data was revealed to the modelers and used as an external validation set to assess the performance of other types of models.

2.2 Machine Learning Methods

We compare the performance of a number of different linear and non-linear machine learning methods, namely Support Vector regression [13], linear ridge regression [14], ensembles of decision trees (Random Forests) [15]. For a description of these well known techniques, we refer to the cited references. In case of the Random Forest method [15] it is worth noting that we use a modified version of the original algorithm, where trees are constructed without bagging or bootstrapping, and pruning of individual trees is done using a CART-style trade-off of error and size. The predictive variance is calculated by averaging the variance of the predictions form the different trees in the forest and the mean estimated variance from training points found at each tree leaf.

The method that turned out to be most powerful for this application is Gaussian Process (GP) regression [6, 2]. The origin of this method lies in Bayesian statistics, where all quantities and parameters involved in model building are treated as random variables. Thus, the prediction of the model is a probability distribution for the quantity of interest. This, in turn, provides the error bars that are important to allow the user to distinguish between predictions that are inside and outside the model's domain of applicability (where model predictions can be assumed to be correct).

3 Results & Discussion

3.1 Choice of Molecular Descriptors

We chose the Dragon descriptors by Todeschini et al. [16]. They include 20 different types of descriptors and are organized in blocks. We assessed the contributions of individual descriptors by training models on subsets of the training data. The resulting ranking is plausible in the context of lipophilicity: The 30 interpretable descriptors given highest weight include counts of donor atoms for H-bonds (N and O), H attached to hetero atom, hydroxyl groups etc. and continuous quantities like the topological polar surface area, molar refractivity, hydrophilic factor and various measures of size. Using only the 50 most important descriptors, we could achieve almost as high accuracy as with the full set, but in some experiments, the associated error predictions turned out too optimistic. Including whole Dragon blocks containing important descriptors leads to both accurate predictions and accurate error estimations. In this study, we used the full Dragon blocks 1, 2, 6, 9, 12, 15, 16, 17, 18, and 20.

3.2 Overall Accuracy: Public Data

Public data Physprop/Beilstein	MAE	RMSE	% ± 1
Gaussian Process	0.38	0.66	92.6
Linear ridge regression	0.59	0.89	84.4
Support Vector Machine	0.40	0.71	91.8
Random Forest	0.52	0.82	87.6
ACDLabs v9	0.43	0.90	89.2
Wskowwin v1.41	0.25	0.90	91.6
AdmetPredictor v1.2.3	0.65	1.32	86.9
QikProp v2.2	0.76	1.23	79.6
baseline: predict mean log P	1.68	2.24	40.7

Table 1. Results when predicting log P on public data (dataset 3). Models are evaluated based on the mean absolute error (MAE), the root mean squared error (RMSE), and the percentage of compounds where the prediction is within one log unit of the true value (% ± 1). For the machine learning methods, the results listed are from 10 times repeated 2-fold cross-validation

Gaussian Processes, Support Vector regression, Random Forest models and the the three best performing commercial tools Wskowwin v1.41, AdmetPredictor v1.2.3 and ACDLabs v9 all perform well on the public set of data, see Table 1. Yet, all four commercial tools produce a number of very "far off" predictions (outliers). In the case of ACDLabs v9 and Wskowwin v1.41 there is a small (less than 10) number of extreme outliers where the log P deviates by more than 10 orders of magnitude. For several dozens of compounds the predicted values are too high by two or three orders of magnitude. Still, the overall performance of both ACDLabs v9 and Wskowwin v1.41 is good, which is also reflected in the low MAE and RMSE. Both AdmetPredictor v1.2.3 and QikProp v2.2 do not exhibit extreme outliers (> ten orders of magnitude). They do, however, overestimate log P for several hundreds of compounds by two or three orders of magnitude, resulting in lower overall performance.

Measurements from the Beilstein and Physprop databases are included in the training data of all four commercial tools. Consequently, it is not surprising that these models perform very well when evaluated on the public set of data. In our own cross-validation procedure, models are only evaluated on compounds that have not been included in the respective set of training data. However, training and test set are sampled from the same distribution across different compound classes. Even with random sampling, the underlying distribution will be preserved, i.e., performance estimates will only be unbiased for equal test and training distributions. When applying models in drug discovery and development, this assumption does generally not hold: In new projects, new compound classes will be investigated. It is therefore desirable to evaluate models with new sets of data. Ideally, the evaluation is done by an independent evaluating team [12]. Such an evaluation is described in the following section.

3.3 Overall Accuracy: In-house Data

In-house cross-validation	MAE	RMSE	% ± 1
Gaussian Process	0.41	0.66	90.7
Linear ridge regression	0.53	0.96	88.3
Support Vector Machine	0.44	0.70	89.8
Random Forest	0.55	0.80	84.4
ACDLabs v9	1.41	1.90	46.6
baseline: predict mean $\log D_7$	1.13	1.47	53.4

In-house blind test	MAE	RMSE	% ± 1
Gaussian Process	0.60	0.82	81.2
Linear ridge regression	0.60	0.83	82.2
Support Vector Machine	0.58	0.81	81.6
Random Forest	0.74	1.00	74.8
ACDLabs v9	1.40	1.79	44.2
baseline: predict mean $\log D_7$	1.17	1.51	51.7

Table 2. Results achieved when predicting $\log D_7$ on in-house data of Bayer Schering Pharma. Models are evaluated based on the mean absolute error (MAE), the root mean squared error (RMSE), and the percentage of compounds where the prediction is within one log unit of the true value (% ± 1)

The Gaussian Process model, Support Vector regression and linear ridge regression yielded good results, when when applied to Bayer Schering Pharma in-house measurements of $\log D_7$ in 2-fold cross-validation, see Table 2. The Gaussian Process model performed best (90.7% ± 1), and was then evaluated in a blind test at at Bayer Schering Pharma, using 7013 new measurements from the last months, see Figure 1 and Table 2. Out of the commercial tools that were available to us (see Table 1) only ACDLabs can calculate $\log D_7$. It predicts only 44.2% of the compounds correctly within one log unit.[6] Our own GP model predicts 81.2% of all compounds correctly within one order of magnitude. These are very promising results, considering that the evaluation was done in a blind

[6] One has to keep in mind that the model in ACDLabs has been trained using shake-flask measurements. The in-house measurements used in this study were performed using HPLC with a hydrocarbon stationary phase and a polar eluent.

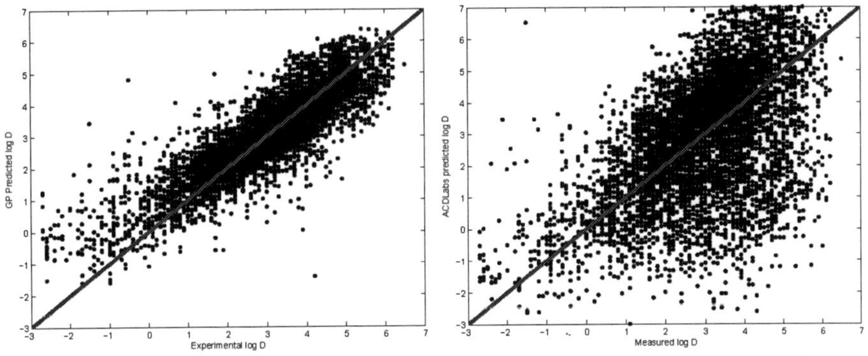

(a) GP (in-house blind test) (b) ACD (in-house blind test)

Fig. 1. Scatter-plots for predicting $\log D_7$ on the in-house blind test data

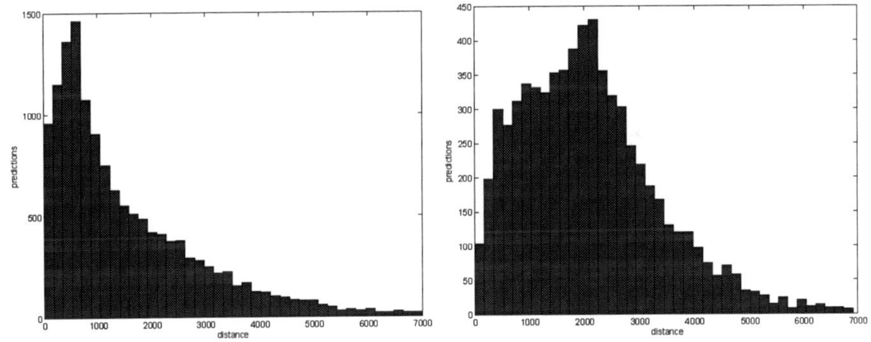

(a) In-house cross-validation (b) In-house blind test

Fig. 2. Histograms of Mahalanobis distances from each compound to the closest compound in the respective training set. Distances for the cross validated in-house setup a) were calculated for the training/validation-split of one arbitrarily chosen cross-validation run.

test setting, using a set of new compounds with a different distribution across the various compound classes. Later, the experimental log D_7 values were made available to the modelers. Support Vector regression, linear ridge regression and Random Forest models were then evaluated as well, using the former blind-test set as an external validation set. For all four types of models tested, the performance on the blind/external set of data is somewhat lower than initially estimated in cross validation. Typical symptoms of overfitting to the training data (e.g. too many support vectors in a support vector regression model) could not be observed. However, in a typical cross-validation run on the in-house set of data, 50 % of the compounds have a nearest neighbor at a descriptor based Mahalanobis distance smaller than 1100 units, see Figure 2. In the blind test set, less than 25 % of the compounds have such close neighbors. This is taken as an indicator of the structural differences between the sets. Therefore it should be possible to achieve higher performance by focusing on compounds that are clearly inside the domain of applicability of the respective model. Results of experiments in this direction are presented in Sec. 3.5.

3.4 Individual Error Estimation for Interactive Use

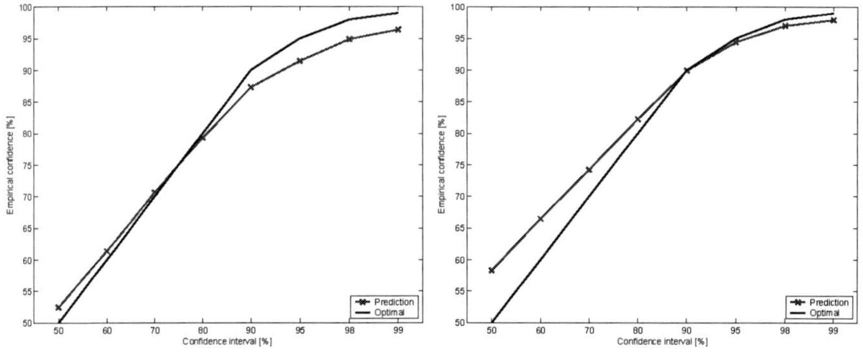

(a) GP (in-house blind test) (b) SVM (in-house external validation)

Fig. 3. Predicted error bars from the GPlogD model exhibit close to ideal statistical properties. Support Vector regression with descriptor based Mahalanobis distance for error estimation is slightly underconfident.

A common heuristic to obtain error estimates for a predictive model is to fit a function (e.g., linear or exponential) to the relationship between the distance of compounds to the training data and observed errors, and use it to generate an error prediction for each prediction the model makes. But how does the user know, what an error prediction of, e.g., 0.6 log units really means? In how many cases does one have to expect the error to be larger than the predicted error? How much larger can errors turn out?

In the natural sciences, uncertainty in measurements or predictions is most commonly quantified using error bars. Gaussian measurement noise is assumed and it follows from the definition, that the true value has to be in the interval $\bar{f} \pm \sigma$ with 68 % confidence, and in the interval $\bar{f} \pm 2\sigma$ with 95% confidence, etc. When evaluating a prediction method that produces error bars, one has to check, whether these error bars have the correct statistical properties: Are really 68 % of the true values within one σ of the predictions? This question should not only be answered for 1σ, 2σ etc., but also for a number of confidence intervals in between. Such an evaluation is presented in in Figure 3.

Predicted error bars of all four models exhibit the correct statistical properties, with the GPlogD error predictions being closest to the ideal distribution. We conclude that it is possible to represent the uncertainty in predictions in an intuitively understandable way, building on the users' knowledge about the notion of error bars.

3.5 Increasing Accuracy by focusing on the Domain of Applicability

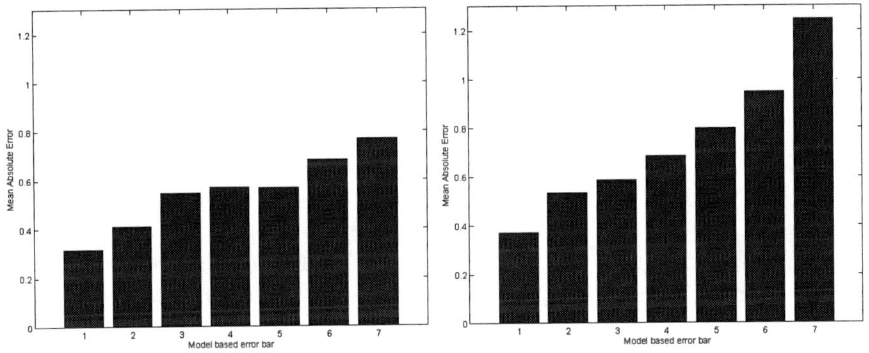

(a) GPlogD (b) Random Forest

Fig. 4. Exploiting model based error bars, the mean absolute error can be reduced significantly. The in-house validation set is binned by predicted errors so that each of the seven bins contains 1000 compounds.

distance (av. in bin)	448	1021	1534	1986	2428	3057	5256
MAE (RR)	0.43	0.50	0.52	0.62	0.66	0.73	0.73
MAE (SVM)	0.35	0.44	0.50	0.61	0.65	0.73	0.79

Table 3. Using distance based error estimation for the models that do not readily supply error bars, the mean absolute error can also be reduced significantly.

In the previous section we showed that Gaussian Processes predict error bars with close to ideal statistical properties. Starting from this promising result, we can now check how much we can really gain by taking these predicted error bars into account. To answer this question, we proceeded as follows: We split the whole ~ 7000 compound external validation set into seven bins, such that the first bin contains the 1000 predictions with the highest confidence (smallest error bar) and so on. In Figure 4 we plotted the mean absolute error achieved in each bin for the Gaussian Process models and Random Forests. For the Support Vector Machine and the linear ridge regression, results are generated in the same way, but based on error estimation using the descriptor based Mahalanobis distance to the closest compound in the training set, can be found in Table 3.

Across all algorithms for modelling and error estimation we observe a clear trend: In bins with smaller predicted error bars we actually find smaller errors. This trend is most pronounced for the Gaussian Process model, where the mean absolute error for the 1000 most confident predictions is only 0.32 log units, see Figure 4. Calculating the mean absolute error for the 3000 most confidently predicted compounds, the result is 0.42 log units, that is almost identical to the 0.41 estimated in cross validation on the in-house data. We conclude that statistics for the blind test data turn out as good as previously estimated in cross validation on the first batch of in-house data if the predicted error bars are used to focus on predictions inside the domain of applicability of the model.

4 Summary

We evaluated four different types of machine learning models for predicting the lipophilicity of public and in-house data. The accuracy of most models compares favorably with four commercial tools on the public data. On the (much more important) in-house data a large performance gain could be achieved over the commercial tools. These positive results are re-confirmed by a blind evaluation on a large set of measurements from new drug discovery projects at Bayer Schering Pharma.

The best performing model, that is based on Gaussian Process regression, is not only capable of making accurate predictions, but can also be used as a fully automatic adaptable tool: Using Bayesian model selection criteria allows for re-training without user intervention whenever new data becomes available.

Predicted error bars form the GPlogD model are shown to exhibit close to ideal statistical properties and therefore allow to represent the uncertainty in predictions in an intuitively understandable way. Furthermore, using predicted error bars allows to achieve substantial performance gains by focussing on compounds for which reliable predictions can be made.

Acknowledgments

The authors gratefully acknowledge DFG grant MU 987/4-1 and partial support from the PASCAL Network of Excellence (EU #506778). We thank Vincent

Schütz and Carsten Jahn for maintaining the PCADMET database and Gilles Blanchard for implementing the Random Forest method as part of our machine learning toolbox.

References

1. Schroeter, T., Schwaighofer, A., Mika, S., Laak, A.T., Suelzle, D., Ganzer, U., Heinrich, N., Müller, K.R.: Estimating the domain of applicability for machine learning qsar rmodels: A study on aqueous solubility of drug discovery molecules. Journal of Computer Aided Molecular Design (2007)
2. Schwaighofer, A., Schroeter, T., Mika, S., Laub, J., ter Laak, A., Sülzle, D., Ganzer, U., Heinrich, N., Müller, K.R.: Accurate solubility prediction with error bars for electrolytes: A machine learning approach. Journal of Chemical Information and Modelling **47**(2) (2007) 407–424
3. Bruneau, P., McElroy, N.R.: Generalized fragment-substructure based property prediction method. J. Chem. Inf. Model. **46** (2006) 1379–1387
4. Schroeter, T., Schwaighofer, A., Mika, S., ter Laak, A., Sülzle, D., Ganzer, U., Heinrich, N., Müller, K.R.: Predicting lipophilicity of drug discovery molecules using gaussian process models. ChemMedChem (2007)
5. Schroeter, T., Schwaighofer, A., Mika, S., Laak, A.T., Suelzle, D., Ganzer, U., Heinrich, N., Müller, K.R.: Machine learning models for lipophilicity and their domain of applicability. Mol. Pharm. **4**(4) (2007) 524–538
6. Rasmussen, C.E., Williams, C.K.: Gaussian Processes for Machine Learning. MIT Press (2005)
7. Netzeva, T.I., Worth, A.P., Aldenberg, T., Benigni, R., Cronin, M.T., Gramatica, P., Jaworska, J.S., Kahn, S., Klopman, G., Marchant, C.A., Myatt, G., Nikolova-Jeliazkova, N., Patlewicz, G.Y., Perkins, R., Roberts, D.W., Schultz, T.W., Stanton, D.T., van de Sandt, J.J., Tong, W., Veith, G., Yang, C.: Current status of methods for defining the applicability domain of (quantitative) structure-activity relationships. Alternatives to Laboratory Animals **33**(2) (2005) 1–19
8. Physical/Chemical Property Database (PHYSPROP), Syracuse, NY, USA
9. Beilstein CrossFire Database, San Ramon, CA, USA
10. Sadowski, J., Schwab, C., Gasteiger, J.: Corina v3.1, Erlangen, Germany
11. Todeschini, R., Consonni, V., Mauri, A., Pavan, M.: DRAGON v1.2, Milano, Italy
12. Müller, K.R., Rätsch, G., Sonnenburg, S., Mika, S., Grimm, M., Heinrich, N.: Classifying 'drug-likeness' with kernel-based learning methods. J. Chem. Inf. Model **45** (2005) 249–253
13. Smola, A., Schölkopf, B.: A tutorial on support vector regression. Statistics and Computing (2001) Forthcoming.
14. Hastie, T., Tibshirani, R., Friedman, J.: The Elements of Statistical Learning: data mining, inference and prediction. Springer series in statistics. Springer, New York, N.Y. (2001)
15. Breiman, L.: Random forests. Machine Learning **45** (2001) 5–32
16. Todeschini, R., Consonni, V.: Handbook of Molecular Descriptors. John Wiley & Sons, Ltd. (2000)

Ab Initio potential grid based docking: From High Performance Computing to *In Silico* Screening

Marc R. de Jonge[1], H. Maarten Vinkers[1], Joop H. van Lenthe[2], Frits Daeyaert[1], Ian J. Bush[3], Huub J. J. van Dam[4], Paul Sherwood[4], Martyn F. Guest[5]

[1] MolMo Services BVBA, Campus Blairon 424, B-2300 Turnhout, Belgium
corresponding author: E-mail marc@molmo.be
[2] Theoretical Chemistry Group, Utrecht University, Padualaan 8, 3584 CH, Utrecht, The Netherlands
[3] Science and Technology Facilities Council, Daresbury Laboratory, CSE, ARCG, Warrington, WA4 4AD, United Kingdom
[4] Science and Technology Facilities Council, Daresbury Laboratory, CSE, CCG, Warrington, WA4 4AD, United Kingdom
[5] Advanced Research Computing Division (ARCCA), Cardiff University, 39-41 Park Place, Cardiff CF10 3BB, Wales United Kingdom

Abstract

We present a new and completely parallel method for protein ligand docking. The potential of the docking target structure is obtained directly from the electron density derived through an *ab initio* computation. A large subregion of the crystal structure of Isocitrate Lyase, was selected as docking target. To allow the full *ab initio* treatment of this region special care was taken to assign optimal basis functions. The electrostatic potential is tested by docking a small charged molecule (succinate) into the binding site. The *ab initio* grid yields a superior result by producing the best binding orientation and position, and by recognizing it as the best. In contrast the same docking procedure, but using a classical point-charge based potential, produces a number of additional incorrect binding poses, and does not recognize the correct pose as the best solution.

1 Introduction

1.1 Docking

Molecular docking is one of the best established computational methods to predict the affinity of small molecules towards a target enzyme or protein structure. The basic tenet of the method is that for a biologically relevant effect the molecule has to find an energetically favourable binding geometry, orientation, and position on or inside the target. This goes both for classical 'lock-and-key' interactions[1], where the geometry for the enzyme is independent of the substrate, and for the related 'induced-fit' type interactions[2], where the lock geometry is forced by the substrate or an allosteric effector. The docking method has become a widely used method for rational drug design because, theoretically at least, it allows the evaluation of completely novel compounds against any target system for which the 3D structure is known in sufficient detail.

As drugs are very often enzyme inhibitors that need to bind to their target in a specific way, pharmaceutical research has produced many computational methods for determining this optimal geometry of the complex formed by inhibitor and target enzyme in its search of new drugs. In essence all docking methods generate a score

CP940, *CompLife 2007, 3rd International Symposium on Computational Life Science*
edited by A. Siebes, M. Berthold, R. Glen, and A. Feelders

for the interaction between the systems by a function that provides a simplified representation of the target or of some properties of the binding cavity inside the target, where this target is based on an X-ray, or occasionally NMR, derived structure of the enzyme. The scores use either a molecular mechanics (also know as force-field) based classical model for the interaction energy (see for instance [3,4,5,6]) or an empirical model that assigns a numerical score to the different recognized fragments of the ligand (for instance [7,8]) or the interacting atom types [9] and deriving a quality of fit measure from known protein-substrate and protein-inhibitor complexes.

Many improvements have been made to these docking methods during the past years, especially with respect to the global optimization strategies by which positions, orientations, and internal geometries of the docked ligands arc generated, and with respect to the level of detail with wich molecules or fragments are parameterized. However, so far we find one aspect of optimization curiously lacking: Even the most elaborate physical scoring functions determine electrostatic potentials from a point-charge model for the enzyme interior, also for the very closest contacts between enzyme and inhibitor. Obviously in the mid-1990's deriving the potential grid from such a model was the maximum that was practically achievable for rational drug design purposes. But, with current hardware we think this particular simplification of the problem may no longer be warranted. Therefore we will develop and apply a new scoring function for docking that uses a full *ab initio* computed electrostatic potential for a more interaction energy estimate as docking score. By this we hope to avoid some of the artefacts introduced by the classical approach, especially the large number of computed docking poses that needs to be taken into account to avoid missing the correct one.

1.2 Potential grid docking

In practical docking problems the interaction energies between a few thousand protein atoms of the target and up to a hundred atoms of the inhibitor need to be evaluated. This has to be repeated for each geometry-orientation point needed in the search algorithm. In a typical pharmaceutical in silico screening study this has to be subsequently repeated for many thousands of molecules. The number of mathematical operations for a single energy is $O(10^6)$ ($N_{target}.N_{ligand}$), including a square root for the distance. The total number of single-point energy computations will therefore easily reach $O(10^{11})$. Some approximations are therefore necessary because even the simple expression of an interatomic interaction using only Coulomb and Van der Waals interactions (1) is expensive to compute.

$$E_{i,j}=C\frac{q_i q_j}{r_{i,j}}-\frac{B_{i,j}}{r_{i,j}^6}+\frac{A_{i,j}}{r_{i,j}^{12}} \qquad (1)$$

In (1) the $A_{i,j}$ and $B_{i,j}$ describe the Van der Waals parameters for the interaction between target atom i and ligand atom j (See for instance [5]). They can be approximated by a geometric mean of coefficients depending on one atom type only:

$$A_{i,j}\simeq\sqrt{a_i}\sqrt{a_j} \qquad (2)$$

The grid based algorithms can then reduce the number of operations through precomputing the protein part of the potentials. As the normal situation in docking is that the target enzyme has $O(10^4)$ atoms whereas the ligands have $O(10^2)$ atoms, the reduction of computation is in practice better than the reduction from N^2 to N dependence that the simple i-j interaction would suggest. The three precomputed Coulomb and Van der Waals dispersion and repulsion grids consist of the sum over all target atoms i of (1):

$$V_{cou\ k,l,m} = C \sum_{i=1}^{N_{target}} \frac{q_i}{r_{i,klm}}$$
$$V_{disp\ k,l,m} \simeq \sum_{i=1}^{N_{target}} \frac{\sqrt{b_i}}{r_{i,klm}^{6}} \qquad (3)$$
$$V_{rep\ k,l,m} \simeq \sum_{i=1}^{N_{target}} \frac{\sqrt{a_i}}{r_{i,klm}^{12}}$$

(Another approach for grid generation, with a separate grid for each target atom type 'i', is used in [3], this improves accuracy but requires much more memory).

The interaction energy of a ligand atom j located at grid cell k,l,m with all target atoms is then as simple linear combination of three grid values:

$$E_j \simeq V_{cou\ k,l,m} q_j - V_{disp\ k,l,m} \sqrt{b_j} + V_{rep\ k,l,m} \sqrt{a_j} \qquad (4)$$

In this study we will replace the V_{cou} grid, as above derived from the partial point charges of the target atoms, with an *ab initio* computed electrostatic potential. We will compare the resulting energy scores and docking poses of a known bound inhibitor in the *ab inito* and the point charge based potentials.

1.3 Test target systems

For testing the approach we will use a simple and a more complicated system: a single water molecule and the strongly polar, Mg^{2+} containing, binding site of *mycobacterium tuberculosis* Isocitrate Lyase (ICL), as described in Protein Databank [10] entries 1F61, 1F8I and 1F8M [11]. Hydrogens were added to the crystal structure, and the geometry was optimized by means of a molecular mechanics relaxation of the internal strain.

2 Methods

2.1 Existing flexible ligand - rigid enzyme docking

We will use a genetic algorithm for flexible ligand docking as implemented in GACS[12] for the docking experiments. The GACS program takes precomputed grids from any source as input for its search for optimal binding orientation and geometry. The program has been used with both 'soft' (4-8) or 'hard' (6-12) Lennard-Jones potentials (Equation 3). For this study we will only use the 6-12 potential; as we are docking a known bound ligand back into its binding site there is no need for the 'fuzziness' of the soft potential that is often useful in poorly constrained or flexible targets. The docking of ligands is performed by means of an evolutionary optimization[13] where the fitness of the orientation is the grid-based energy of the compound, incremented with an internal strain component (resulting from the torsions and out of plane forces of the flexible ligand). The real-valued chromosomes that describe a docking orientation inthe evolution thus consist of 6+n real numbers: three Euler angles, a 3D displacement vector, and n torsion/out of plane angles. This algorithm has been employed, with classical electrostatics, in the past in many docking studies and is robust with respect to protein and ligand electronic properties.

2.2 GAMESS-UK grid generation

The most important reason *ab initio* grids are normally not used for docking is the sheer size of the computational problem. In order to produce a good electronic structure a very large number of atoms has to be taken into account: The size of the enzyme test system requires a careful setup of the system; the complete enzyme of ICL has about 18000 atoms, but from classical electrostatics models we know that a sub-region with a radius of 16Å or of about 2000 atoms is generally sufficient for docking studies.

In this study we have computed the potential function with GAMESS-UK, a package of *ab inito* progams [14]. The *ab initio* grid generation entails first calculating the wave function and subsequently analyzing that through the generation of potential and electron density grids. The most time consuming part is the calculation of the wave function, This involves an direct implementation [15] of either a DFT (B3LYP [16]) or a Hartree-Fock algorithm [17]. The 2-electron integrals are calculated as needed, multiplied by density matrix elements (P in (5)) and added into the Fock matrix elements. The most common approach to parallelize this problem is through replicated data parallelism: The same program runs on every node of the system, each maintaining an identical copy of the data structures. The computational load is divided across the processors, and inter-processor communication phases are introduced to share the results so that all processors have the same data at the start of the next phase. It involves the contraction of blocks of 4-index two-electron integrals with the density matrix P (Eq. 5).

$$F_{\mu\nu}=H^{0}_{\mu\nu}+P_{\mu\nu}\sum_{\lambda\sigma}\langle\mu\nu|\sigma\lambda\rangle-\frac{1}{2}\langle\mu\lambda|\sigma\nu\rangle \qquad (5)$$

For each block of integrals, with a range of indices $\mu\nu\sigma\lambda$, a number of elements of P is accessed and the result is summed into the appropriate elements of F. All nodes can work independently until the construction is complete, at which point a global summation of contributions to F is required. This approach towards parallelization is readily implemented and has been in use since 1986 [18].

However, as merely parallelizing the Fock-matrix builder being the most time consuming part of the calculation, does not allow scaling beyond 16 to 32 processors, we have progressively parallelized more of the code by introducing the Global Array [19] and Parallel Eigenvalue solver PeIGS [20] tools developed at Pacific Northwest National Laboratories. These allow similarity transforms and diagonalization to be executed in parallel. Scalability limitations now arise from the structure of the underlying serial program, specifically its heavy use of scratch file I/O and the costs of maintaining the replicated data structures. In some phases of the calculation, matrices which are temporarily distributed across the nodes (for example, in the parallel matrix multiply) must be re-replicated for the next phase of the calculation. In the current calculations the SGI ALTIX machine at SARA [21] was employed. The scaling behaviour of the code is shown to be perfect up to 30 processors (Fig. 1). Beyond that, the scaling deteriorates; this is probably due to the architecture of the machine (which employs 2 processors per node). The memory requirement in replicated data parallel programs is the same per processor as that of the serial code. On the platform used this results in a limit of about 10000 basis functions, since a density matrix and a Fock matrix have to be stored and manipulated. A shared memory implementation of the parallel code is currently being developed at the STFC Daresbury Laboratory [22] to extend this upper limit [23]. This facility has a shared memory architecture machine that needs to store only one density and Fock matrix per node with 32 processors. An illustration of the performance of the new code in this environment, and for some larger cutouts of the enzyme, is given in figure 2. The poor result of the 20 Å duplicated memory run was due to the fact that only 12 cpu's of the node are effectively used because of memory limitations. After determining the HF solution the final parallel computation of the grid values is fast and efficient.

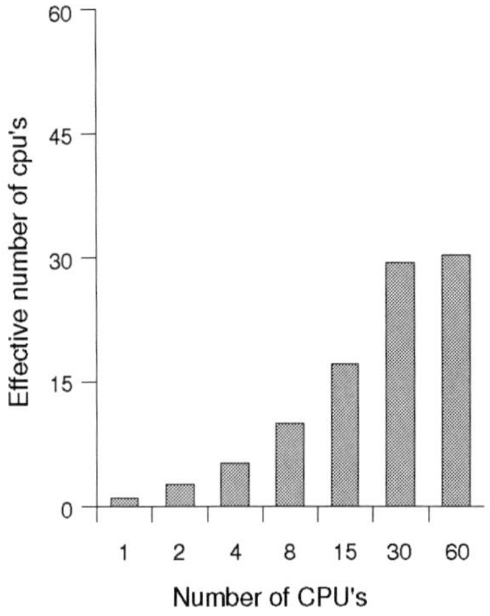

Figure 1: Scaling of a Hartree-Fock calculation of a 12 Å radius region of ICL: 737 atoms and 4500 orbitals.

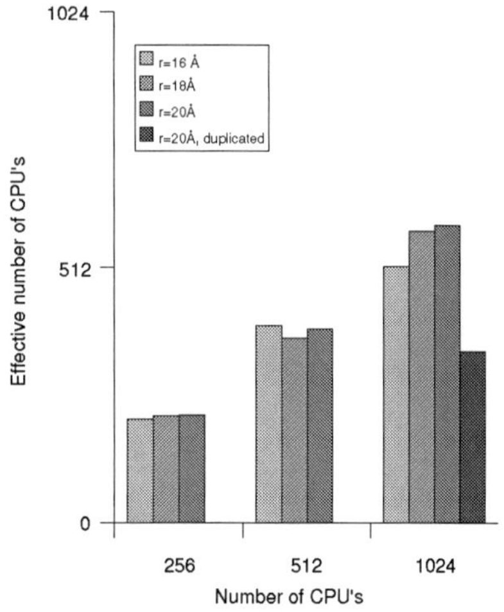

Figure 2: Scaling of Hartree-Fock calculation of 16, 18, and 20 Å radius subsets of the ICL enzyme with shared memory code.

Because of above mentioned practical constraints the number of basis functions in the quantumchemical computation needs to be limited. Obviously for the tests on the single water molecule a TZVP basis set was used without problems. But the protein structure of ICL has to be prepared with some care: After protonating the crystal structure, we first relaxed the whole system by means of 10000 steps of conjugate gradient minimization with the protein backbone atoms constrained to their original position by means of a quadratic displacement penalty of 20 kCal mol^{-1} $Å^{-2}$. Subsequently a spherical subvolume is defined by removing (and capping whole) residues that are more than 16 Å away from the Mg^{2+} ion located in the active site.

It is this spherical volume that is modeled in GAMESS-UK. The number of atoms is still so large that not all atoms can be treated at maximum number of basis functions. We defined concentric shells with different basis functions, from TZVP in the central region, through 6-31g at 6 Å, to sto3g starting at 12 Å as illustrated in Figure 3. This shell structure reduces the size of the problem to less than 2000 atoms with 10000 basis functions. The grid cells for the potential calculation are set to 0.2 Å in all directions, the grid size is 121x121x121 cells, centered around the Mg^{2+} ion (the potentials in this grid are computed from the whole 16Å volume).

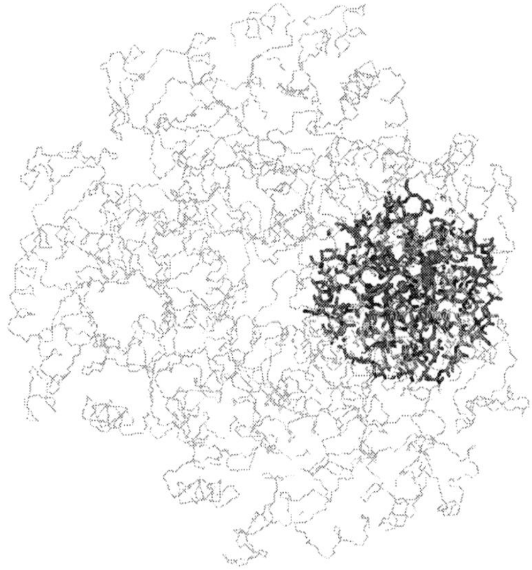

Figure 3: Whole ICL enzyme in light grey, and the 16 Å region with its different basis sets colour coded. The pink atom is the Mg^{2+} ion of the catalytic site, the grey atoms next to it are from the succinate reaction product.

3 Results and discussion

3.1 Potential field of a water molecule

To illustrate the effect of the different ways of computing an electrostatic potential we performed a small scale experiment on a single water molecule. Figure 4a shows the forcefield partial charge based electrostatic potential, and Figure 4b the same derived from NBO computed partial charges [24] from an *ab initio* HF . It is clear that the two methods, both using point charges produce very similar potentials (This is not surprising because the forcefield we used is designed to reproduce the NBO partial charges). In contrast the electrostatic potential computed directly from the QM results in Figure 4c (i.e. without projection onto point charges), is notably different. Not only inside the Van der Waals radius of the atoms, but also outside up to a distance that is very relevant for docking. Even in a 'toy' problem like this, the difference between a point-cloud description and a volume charge description is significant. This difference between the two point-charge derived potentials on one hand and the electron density based potential on the other, at 3 Å from the oxygen atom, is about 4 kCal/Mol (at the positions indicated by the black arrows in Figure 4).

The reason for this, unexpectedly large, difference lies in the fact that electrons of the atoms extend further than the Van der Waals radius, which means that, for the close contacts that are important in docking, the implicit assumption that all partial charge is contained within this volume is not entirely correct. Another effect is that the actual shape of the electronic structure is not really spherically symmetric (as assumed in the point charge potentials 4-a and 4-b). This latter effect also explains the relatively good potential for the four-point TIP4P [25] water model in Fig. 4d which is designed for, and commonly used in, molecular dynamics simulation. This water model has the point charge offset by 0.15 Å from the oxygen nucleus (towards the

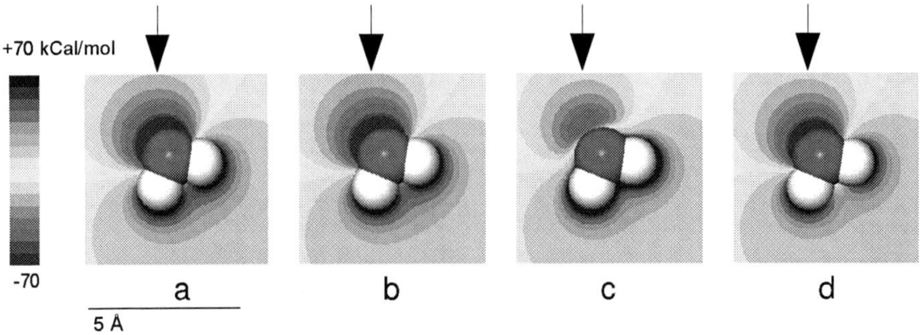

Figure 4: Electrostatic potential of a water molecule. a: Molecular Mechanics point charges, b: ab initio (RHF with NBO) point charges, c: ab initio electron density based, d: TIP4P (H_2O+dummy charge) water.

hydrogens). As a result its coulomb potential shows a better match with the *ab initio* potential at the arrow position, although it still resembles the NBO and forcefield point charge potentials closer to the oxygen atom.

3.2 Potential field of the Isocitrate Lyase enzyme binding site

The active site of ICL can be treated in the same way as the water molecule above, but as the Mg^{2+} ion is not balanced by a ligand (for docking the potential of the unliganded binding region is required) a net charge exists in this region, and consequently a net positive electrostatic potential is found everywhere near the central region. In Figure 5 the potentials computed from point-charges (a) and from the electron density (b) are shown. The red atoms in both binding sites are the oxygens of negatively charged carboxylate groups of glyoxylate and succinate in their crystal structure positions. The potential was computed without the succinate ligand, it is therefore promising for docking that the electrostatic potential shows local maxima at the position of its negatively charged carboxylate groups. There is quite a large difference in magnitude of the electrostatic potential visible, especially in close contact with the enzyme atoms. Another important feature is the generally narrower appearance of the potential extrema in the *ab initio* grid. This suggests that the energy landscape in which is scanned during the docking will have different optimal positions in the two approaches, and less freedom for (mis-)placement of compounds in the *ab initio* grid.

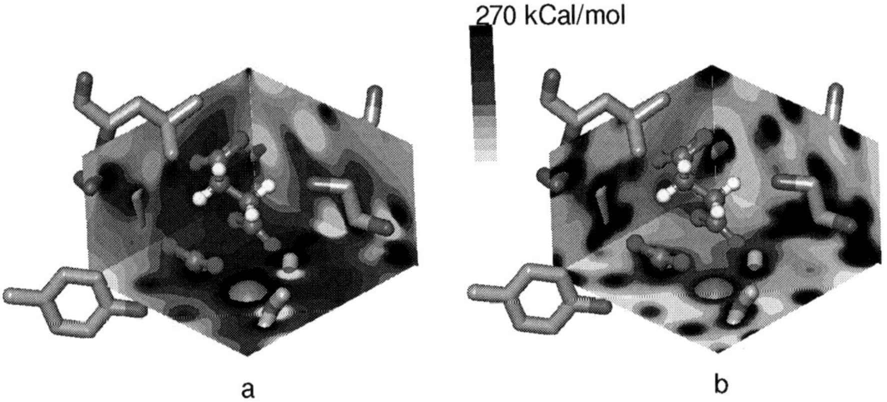

Figure 5: Three sections through the electrostatic potential of the central region of the ICL model. The point charge based model is shown in a, the ab initio computed potential in b. The Mg^{2+} ion, some relevant protein atoms, as well as the glyoxylate and succinate products from isocitrate lysis are shown.

3.3 Docking with the two types of electrostatic potentials

Docking of one of the products of isocitrate lysis, succinate, into the binding site of the enzyme is possible with both potential grids. Figure 6 makes clear however that the two approaches are not equivalent. The compound with the dark green carbon atoms indicate the predicted best binding orientations, in Figure 6a for the point charges based potential, and in Figure 6b for the electron density based potential. An important difference is the shift of both carboxylate groups in the point charge docking result: A hydrogen bond with the coordinated water molecule next to the pink Mg^{2+} ion is lost (arrow a-1), in favour of a 'promiscuous' coulomb interaction within the not very discriminating wide positive potential region (arrow a-2). In the electron density based field the docked ligand retains the interactions that are expected from the crystal structure (a hydrogen bond with a structural water molecule at b-1 and with an arginine side chain at b-3). The point charge based docking yields three more local optima with incorrect docking poses, with a maximum docking energy difference of 13 kCal/mol. The correct orientation is also found, but scores about 2 kCal/mol worse than the lowest energy one*. The electron density grid yields only one optimal binding mode. It should be noted that, when docking small polar molecules, the number of binding orientations found is generally far more than the crystal structures indicate. The fact that, rather than four, only the one correct orientation of succinate is identified is an important advantage of docking with this new potential.

* implying 3.5% of this conformation would be presentin a Boltzmann distribution at 300K.

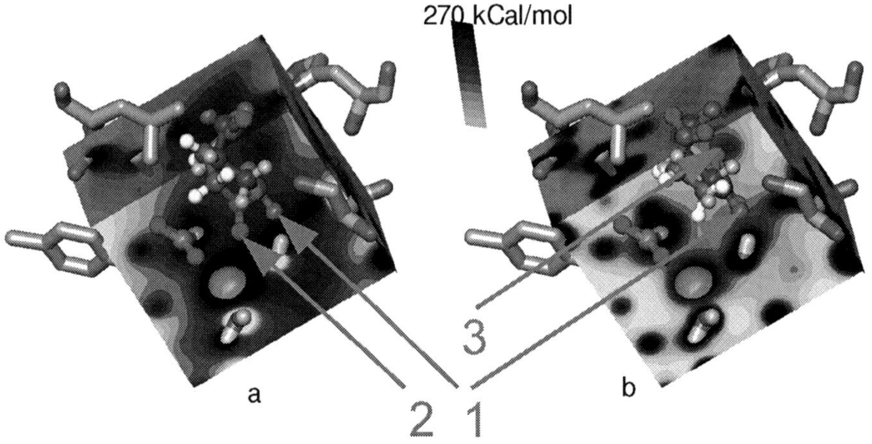

Figure 6: Docking result of succinate (green carbon atoms), a) in point charge potential field and b) in electron density potential field. For reference the dark grey crystal structure orientation of succinate is also shown in both binding sites. See text for the arrows 1,2 and 3.

4 Conclusions

We have demonstrated a parallel an fully *ab initio* approach towards the generation of protein ligand docking grids. The whole computational procedure from grid generation to multiple ligand flexible docking can be performed in parallel. The use of electron density derived potential grids results in a substantial improvement of the quality of docking relative to a model derived from forcefield charges. In *in silico* drug design and *in silico* library screening hundreds of thousands of compounds are screened in parallel, and the amount of work spent on generating the *ab initio* grids is negligible (on the order of several thousands of cpu-hours) relative to the total computational effort involved in the subsequent docking of multiple ligand conformers of the whole library , which can easily amount to an additional 100,000 cpu-hours after computation of the potential grids.

We have also demonstrated that the essential characteristic of the method we propose is the direct generation of the grids for the docking algorithm from electron density without interposing a point-charge layer of approximation, thereby removing the artifacts of that approximation. The difference between direct grids and charge based potential fields is illustrated by using the same electronic structure but with back-projected point charges (eg. through NBO or DMA partial charges), which results in a potential field almost identical to the molecular mechanics model and not similar to the electron density grid based model. The original grid-based docking procedure does not need to be modified in any way to make use of the *ab initio* potentials.

We conclude that the close contacts between target and ligand in docking studies demand a better approximation for the electrostatic potential than the spherically symmetrical approximation commonly used.

Acknowledgments

Part of this work was carried out under the HPC-EUROPE Project (RII3-CT-2003-506079), with the support of the European Community Research Infrastructure Action under the FP6 Structure of the European Research Area Programme.

Part of this work was sponsored by the Stichting Nationale Computerfaciliteiten (NCF) for the use of supercomputer facilities, with financial support from the Netherlands Organization for Scientific Research, NWO.

We thank Luc Koymans for discussion on the protein function and protonation state.

References

1. Fischer, E.: Einfluss der Configuration auf die Wirkung der Enzyme. Berichte der Deutschen chemischen Gesellschaft **27** (1894) 2985-2933
2. Koshland, D.E.: Application of a Theory of Enzyme Specificity to protein Synthesis. Proceedings of the National Academy of Sciences of the United States of America **44** (1958) 98-104
3. Goodford, P.J.: A Computational Procedure for Determining Energetically Favorable Binding Sites on Biologically Important Macromolecules. Journal of Medicinal Chemistry **28** (1985) 849-857
4. Von Itzstein, M., Wu, W.-Y., Kok, G.B., Pegg, M.S., Dyason, J.C., Jin, B., Van Phan, T., Smythe, M.L., White, H.F., Oliver, S.W., Colman, P.M., Varghese, J.N., Ryan, D.M., Woods, J.M., Bethell, R.C., Hotham, V.J., Cameron, J.M., and Penn, C.R.: Rational design of potent sialidase-based inhibitors of influenza virus replication. Nature **363** (1993) 418-423
5. Makino, S., Kuntz, I.D.: Automated Flexible Ligand Docking Method and Its Application for Database Search. Journal of Computational Chemistry **18** (1997) 1812-1825
6. Jones, G., Willett, P., Glen, R.C., Leach, A.R., Taylor, R.: Development and Validation of a Genetic Algorithm for Flexible Docking. Journal of Molecular Biology **267** (1997) 727-748
7. Boehm, H.-J.: The Development of a simple empirical scoring function to estimate the binding constant for a protein-ligand complex. Journal of Computer-Aided Molecular Design **8** (1994) 243-256
8. Wang, R., Liu, L., Lai, L., Tang, Y.: SCORE: A New Empirical Method for Estimating the Binding Affinity of a Protein-Ligand Complex. Journal of Molecular Modeling **4** (1998) 379-394
9. Gohlke, H., Hendlich, M., Klebe, G.: Knowledge-based scoring function to predict protein-ligand interactions. Journal of Molecular Biology **295** (2000) 337-356
10. Berman, H.M., Westbrook, J., Feng, Z., Gilliland, G., Bhat, T.N., Weissig, H.,Shindyalov, I.N., Bourne, P.E.: The Protein Data Bank. Nucleic Acids Research **28** (2000) 235-242
11. Sharma, V., Sharma, S., Zu Bentrup, K.H., McKinney, J.D., Russell, D.G., Jacobs Jr., W.R., Sacchettini, J.C.: Structure of Isocitrate Lyase, a Persistence Factor of Mycobacterium Tuberculosis. Nature Structural Biology **7** (2000) 663-668
12. Daeyaert, F., de Jonge, M.R., Heeres J., Koymans L., Lewi P., Vinkers M.H.,

Janssen P.A.: A pharmacophore docking algorithm and its application to the cross-docking of 18 HIV-NNRTI's in their binding pockets. Proteins **54** (2004) 526-533
13. Holland, J. H.: Adaptation in natural and artificial systems. MIT Press (1992)
14. Guest, M.F., Bush, I. J., van Dam, H.J.J., Sherwood, P., Thomas, J.M.H., van Lenthe, J.H., Havenith, R.W.A., Kendrick, J.: The GAMESS-UK electronic structure package: algorithms, developments and applications. Molecular Physics **103** (2005) 719-747 (see:http://www.cfs.dl.ac.uk/gamess-uk/index.shtml)
15. Almlof, J., Faegri, K. Jr., Korsell, K. J.: Principles for a direct SCF approach to LCAO-MO *Ab-Initio* calculations. Journal of Computational Chemistry **3** (1982) 385-399
16. Stephens, P.J., Devlin, F.J., Chabalowski, C.F., Frisch, M.J.: *Ab Initio* calculation of vibrational absorbtion and circular dichroism spectra using density functional force fields. Journal of Physical Chemistry **98** (1994) 11623-11627
17. Jensen, F.: Introduction to Computational Chemistry, Wiley Chichester New York (1999)
18. Guest, M. F., Harrison, R. J., van Lenthe, J. H., van Corler, L. C. H.: Computational chemistry on the FPS-X64 scientific computers. Theoretica Chimica Acta **71** (1987) 117-148
19. Nieplocha, J., Harrison, R.J., Littlefield, R.J.: Global Arrays: A nonuniform memory access programming model for high-performance computers. Journal of Supercomputing **10** (1996) 169-189
20. Dhillon, I., Fann, G., Parlett, B.: Application of a new algorithm for the symmetric eigenproblem to computational quantum chemistry. Proceedings of the Eighth SIAM Conference on Parallel Processing for Scientific Computing (1997)
21. NWO, Stichting Nationale Computer Facililteiten, SARA, Amsterdam. http://www.sara.nl/userinfo/aster/description/index.html
22. Science & Technology Facilities Council, Daresbury Laboratory. http://www.scitech.ac.uk/About/Struc/Locs/DL/home.aspx
23. Bush, I.J.: New Fortran Features: The Portable Use of Shared Memory Segments. HPCx Technical Report, STFC Daresbury Laboratory (2007)
24. Reed, A.E., Curtiss, L.A., Weinhold, F.: Intermolecular interactions from a natural bond orbital, donor acceptor viewpoint. Chemical Reviews **88** (1988) 899-926
25. Jorgensen, W.L., Chandrasekhar, J., Madura, J.D., Impey, R.W., Klein, M.L., : Comparison of simple potential functions for simulating liquid water. The Journal of Chemical Physics **79** (1983) 926-935

Author Index